KU-361-067

The Behaviour of the Domestic Cat, 2nd Edition

John W.S. Bradshaw

School of Veterinary Sciences
University of Bristol

Rachel A. Casey

School of Veterinary Sciences
University of Bristol

Sarah L. Brown

School of Veterinary Sciences
University of Bristol

www.cabi.org

CABI is a trading name of CAB International

CABI	CABI
Nosworthy Way	38 Chauncey Street
Wallingford	Suite 1002
Oxfordshire OX10 8DE	Boston, MA 02111
UK	USA

Tel: +44 (0)1491 832111	Tel: +1 800 552 3083 (toll free)
Fax: +44 (0)1491 833508	Tel: +1 (0)617 395 4051
E-mail: info@cabi.org	E-mail: cabi-nao@cabi.org
Website: www.cabi.org	

© J.W.S Bradshaw, R.A. Casey and S.L. Brown 2012. All rights reserved. No part of this publication may be reproduced in any form or by any means, electronically, mechanically, by photocopying, recording or otherwise, without the prior permission of the copyright owners.

A catalogue record for this book is available from the British Library, London, UK.

Library of Congress Cataloging-in-Publication Data

Bradshaw, John W. S.
 The behaviour of the domestic cat / John W.S. Bradshaw, School of Veterinary Sciences, University of Bristol, Rachel A. Casey, School of Veterinary Sciences, University of Bristol, Sarah L. Brown, School of Veterinary Sciences, University of Bristol. -- Second edition.
  pages cm
 Includes bibliographical references and index.
 ISBN 978-1-84593-992-2 (pbk.)
 1. Cats--Behavior. I. Casey, Rachel A. II. Brown, Sarah L. III. Title.

SF446.5.B73 2013
636.8--dc23

                                   2012026680

ISBN: 978 1 84593 992 2

Commissioning editor: Rachel Cutts
Editorial assistant: Chris Shire
Production editor: Tracy Head

Typeset by SPi, Pondicherry, India.
Printed and bound in the UK by MPG Books Group.

Contents

Preface to the Second Edition

The behaviour of the domestic cat, *Felis silvestris catus*, has many features that set it apart from other common domestic animals, even its fellow carnivore the dog. Cats seem to have effected a unique and successful compromise between reliance on man and the retention of behaviour patterns that permit an independent existence. During the first decade of the 21st century the cat has become the preferred pet of many owners, valued because it combines an affectionate nature with a degree of independence, as well as more prosaic qualities such as cleanliness and convenience.

This book brings together many disparate studies of the behaviour of *F. catus*, and attempts to marry together the more mechanistic approach (how do patterns of behaviour come about?) with the functional (what are those patterns for, and how do they benefit the animal?). It may seem unusual to ask functional questions about the behaviour of a domestic animal, but although for many individual cats both feeding and breeding are largely controlled by man, many others live much more independently and are not only subject to natural selection but, through their predatory behaviour, are themselves a selection pressure on other species. This book therefore covers cats across all their current lifestyles, from neutered pet to fully independent ('feral').

Unlike the domestic dog, which has seen a resurgence in research interest over the past decade, several aspects of the cat's behaviour have received relatively little attention since the publication of the first edition of this book in 1992. These include the cat's sensory and learning abilities and its repertoire of visual signals. However, major advances have been made in several areas, including cat welfare, social behaviour and domestication, and especially in our understanding and treatment of behavioural disorders, coverage of which has been expanded from one chapter to two.

From a biological perspective, it is often useful to compare one species' behaviour with that of related species, in this case doubly so, since hybrids between the domestic cat and other small felids are beginning to appear as pets. The first chapter deals with these issues, including the genetic relationship between the domestic cat and the rest of its family, the Felidae, and the basis for its domestication. The next two chapters attempt to describe the world in the cat's own terms: what it can see, hear, smell and feel, and how it can bring all this incoming information together so that its surroundings make sense. This is worth establishing at the outset, because their perception of the world, while overlapping with ours, also provides them with information that is not directly accessible to us, especially that provided by the cat's olfactory apparatus. Furthermore, cat intelligence has been portrayed in many ways, from the early behaviourists' picture of a near-automaton to the anthropomorphic sentimentalization of the popular literature; Chapter 3 redresses the balance in favour of modern scientific thinking. Subsequent chapters deal with more specialized aspects of cat behaviour, such as reproduction, development, communication, hunting and feeding. Chapters 8, 9 and 10 are based around a common theme, that of the cat's social abilities. The complex social interactions between cats, many of which we do not yet fully understand, have forced biologists to reinterpret not only the cat's relationship with its conspecifics but also the bond with its owner, and considerations of its welfare.

The concluding chapters, building on those that precede them, lay out our current understanding of 'behaviour problems' in cats, divided into those which are essentially normal behaviours for cats but which owners find undesirable (Chapter 11) and those caused or influenced by pathological or physiological factors (Chapter 12). This framework has been put together primarily by one of us (RAC) from her clinical experience at the University of Bristol's Animal Behaviour Clinic (recognized as a centre for clinical training by the European College of Animal Welfare and Behavioural Medicine).

For those interested in reading further, two recent books can be regarded as complementary to this one. *Wild Cats of the World*, by Mel and Fiona Sunquist, extends the story to the other members of the Felidae. More comprehensive treatments of many of the topics are contained in the multi-author book edited by Dennis Turner and Patrick Bateson, *The Domestic Cat: the Biology of its Behaviour*, which is about to appear in its third edition. Further suggestions can be gleaned from the list of references, which has been deliberately kept short, by using review articles rather than research papers wherever possible, to avoid breaking up the text with large numbers of citations.

Many of our friends and colleagues have helped us with this book, by making suggestions, providing references, allowing us to present their unpublished research and reading drafts of chapters. In no particular order, they are Sandra McCune, Carlos Driscoll, David Macdonald, Michael Mendl, Ian Robinson, Stuart Church, John Allen, Sarah Lowe (Benge), Ruud van den Bos, Charlotte Cameron-Beaumont, Giles Horsfield, Sarah Hall, Sylvia Vandenbussche, Kim Hawkins, Elizabeth Paul, Carri Westgarth, Emily

Blackwell, Kristen Kly, Maryanne Heard, Jenna Kiddie, Anne Seawright and Christine Basse. The responsibility for what appears on the printed pages is, of course, ours alone. Our sincere gratitude also goes to Alan Peters for his ink-and-wash sketches of kittens and cats. Our own research, and that of our students, reported here has been supported by the Waltham Centre for Pet Nutrition, the Biotechnology and Biological Sciences Research Council and the Universities Federation for Animal Welfare. We are especially grateful to Cats Protection for supporting research and academic posts over the past two decades, a contribution that has significantly enhanced scientific understanding of cat behaviour and welfare, and its application in clinical practice.

The Cat: Domestication and Biology

1

Introduction

The domestic cat is by far the most numerous of all the cat family, the Felidae, most of whose other members are listed as approaching or vulnerable to extinction (Macdonald and Loveridge, 2010). While the derivation of the domestic cat from wildcats is far from being completely understood, from a biological point of view relationships between domestic animals and their wild counterparts are crucial to a full understanding of behaviour. This is because it is one of the basic principles of modern ethology that species-specific behaviour patterns contain an inherited component, and therefore can be compared from species to species as if they were a morphological character, like the shape of the skull. While it is possible to study the behaviour of the domestic cat in isolation, we are likely to gain more insight into the origins of many behaviour patterns if we can compare them to those of related undomesticated species. Having said that, behavioural studies of closely related species are sparse, other than the 'big cats', notably the lion.

All living species in today's cat family are thought to descend from an Asian genus *Pseudaelurus* that lived some 11 million years ago. Migrations from Asia to Europe, and to North and then South America, with some later reverse migration, produced the distribution of small cat species that we see today (O'Brien *et al.*, 2008; Table 1.1). Many of these species are easy to tame, and historically have been valued as pest controllers in different parts of the world – for example, the jaguarundi in South America and Pallas' cat in Central Asia (Faure and Kitchener, 2009). However, it is now undisputed that the wildcat *Felis silvestris* is the sole original ancestor of the domestic cat (Faure and Kitchener, 2009), although some breeds have recently been produced by hybridization with other

© J.W.S. Bradshaw, R.A. Casey and S.L. Brown 2012. *The Behaviour of the Domestic Cat*, 2nd Edn (J.W.S. Bradshaw *et al.*)

Table 1.1. Some of the more common members of small cats, including their species names (sometimes all included in one genus, *Felis*, their generic names are frequently revised), and their approximate ranges and habitat preferences (derived from Kitchener, 1991; Sunquist and Sunquist, 2002; and O'Brien and Johnson, 2007). One group probably originated in the New World, and is now largely confined to South America. The other groups have Old World origins. The three lineages not included are Panthera ('big cats'), the lynxes and puma/cheetah/jaguarundi. The only races listed are those of *F. silvestris*, of which the domestic cat is one. T, tameability (from Cameron-Beaumont *et al.*, 2002): high, >50% of individuals studied; med, 25–50%; low, <25%; –, not studied.

Common name	Species	Range: habitat	T
Ocelot lineage New World small cats			
Ocelot	*pardalis*	Arizona to Uruguay: wide range of habitats	High
Margay	*wiedii*	Costa Rica to N. Argentina: mainly forest	High
Tiger cat	*tigrina*	Costa Rica to N. Argentina: forest	–
Geoffroy's cat	*geoffroyi*	Bolivia to Patagonia: woodland, bush	High
Kodkod	*guigna*	Chile: forest	–
Pampas cat	*colocolo*	Ecuador to Patagonia: wide range of habitats	–
Mountain cat	*jacobita*	Andes: steppe	–
Bay cat lineage South Asian small cats			
Bay cat	*badia*	Borneo: tropical forest and rocky scrub	–
Asian golden cat	*temminckii*	Himalayas to S.E. Asia: high forest	–
Marbled cat	*marmorata*	S. Asia: forest	–
Caracal lineage African small to medium cats			
Caracal	*caracal*	India to Southern Africa: wide range of habitats	–
Serval	*serval*	Africa: mainly savannah	–
African golden cat	*aurata*	Senegal to Zaire, Kenya: high rainforest	–
Leopard cat lineage Asian small cats			
Pallas' cat	*manul*	Iran to W. China: steppe, desert	–
Rusty-spotted cat	*rubiginosus*	S. India, Sri Lanka: wide range of habitats	Low
Leopard cat	*bengalensis*	S. and E. Asia: woodland, forest, scrub	None
Fishing cat	*viverranus*	S. and S.E. Asia: near water	Low
Flat-headed cat	*planiceps*	Borneo, Sumatra, Malaysia: forest near water	–

Continued

Table 1.1. Continued.

Common name	Species	Range: habitat	T
Domestic cat lineage Old World small cats			
Jungle cat	*chaus*	Egypt to India: wide range of habitats	Med
Black-footed cat	*nigripes*	Southern Africa: desert, grassland	–
Sand cat	*margarita*	Sahara to Turkestan: desert and semi-desert	High
Wildcat	*silvestris*		
European wildcat	*s. silvestris*	Scotland to Georgia: mainly forest	Low
African wildcat	*s. lybica*	Northern Africa, Western Asia	High
Caffer cat	*s. cafra*	Southern Africa	
Indian desert cat	*s. ornata*	N.W. India to Iran, Kazakhstan: semi-desert	Med
Chinese mountain cat	*(s.) bieti*	N. China, Mongolia: steppe and mountain	–
Domestic cat	*s. catus*	Worldwide: associated with man	

cat species. Previously divided into several species and many races, recent DNA evidence indicates that *F. silvestris* can be divided into six types: the European wildcat *F. s. silvestris*; the African/Arabian wildcat *F. s. lybica*; the Indian desert cat *F. s. ornata*; the South African wildcat *F. s. cafra*; and the Chinese mountain cat *F. bieti* (Driscoll *et al.*, 2007; Fig. 1.1). The domestic cat, formally classified as *F. s. catus*, is derived from *F. s. lybica*.

As a species, *F. silvestris* is highly variable in terms of tractability, almost as variable as the Felidae as a whole, with *lybica* the easiest, and *ornata* intermediate, while *F. s. silvestris* is almost impossible to tame (Cameron-Beaumont *et al.*, 2002). The *lybica* form is the most adaptable to living alongside and within human settlements, supporting the DNA evidence for its being the ancestor of *catus*. Other evidence to support this origin comes from archaeology (see below) and the names used to describe cats, many of which, including 'puss', 'tabby' and the word 'cat' itself, have North African or Middle Eastern origins. Although it has often been suggested that the oriental breeds of cat were originally derived from the light-bodied Indian desert cat *ornata*, DNA analysis of the oriental breeds has so far failed to confirm this idea (Lipinski *et al.*, 2008).

From time to time it has been suggested that other species of *Felis* whose ranges overlap with *lybica* might also have contributed to the modern *catus*, including the jungle cat *F. chaus*. The Egyptians tamed large numbers of jungle cats, and remains of cats with skulls intermediate in size between *chaus* and *lybica* have been found in Egyptian tombs, which may be those of hybrids bred in

Fig. 1.1. The distribution of the races of *Felis silvestris c.*10,000 years ago (YA),
showing the likely locations of initial commensalism (C, 10,000–4000 YA) and
subsequent domestication (D, 4000–2500 YA). Light shading, *F. s. silvestris*;
medium shading, *F. s. ornata*; dark shading, *F. s. lybica*; black, *F. s. bieti*; stippled
shading, *F. s. cafra*.

captivity. In recent times a new breed of cat, the 'Chausie', has been created by
hybridizing *catus* with *F. chaus*. Other new breeds have been created by hybridizing *catus* with non-sympatric species such as the serval (the 'Savannah') and
Geoffroy's cat (the 'Safari'). Perhaps the most widespread is the Bengal, a hybrid
with the Asian leopard cat, an unlikely choice since this species is among the
least tameable of all the Felidae. Generally, the animals of these breeds kept as
pets contain only a small proportion of non-*silvestris* genes; for example, F1–F3
Bengals are generally only kept for breeding, with F4 (6% *bengalensis*) and subsequent generations kept as domestic pets.

Apart from these hybrids, most of the pedigree or 'fancy' breeds, such as the Persian, Abyssinian and Turkish Van, have been produced by selective breeding of *catus*, rather than by cross-breeding with other species of *Felis*. The Siamese, Korat, Burmese, Birman and other 'oriental' breeds have distinctive DNA, indicating that they were derived, several hundred years ago, from domestic *catus* populations in areas from India eastwards where there are no naturally occurring *silvestris* (Lipinski *et al.*, 2008). The Turkish Van, Mau and Angora are offshoots of Near Eastern *catus*, while most other pedigree breeds, including the Persian/Exotic, are recently derived from Northern European stock. Although the breeds are primarily defined by their morphology, part of the process has been a selection for breed-specific behavioural traits, which will be discussed below.

The colour forms of *F. catus* (including most of those in the 'fancy' breeds) have arisen by somatic mutation from the original striped tabby coat of *lybica*. The main variations are the blotched tabby, which has three black stripes on the back and spiral or circular stripes on the flanks; the sex-linked orange tabby (which produces orange males and females, and the heterozygous tortoiseshell and calico females); the plain black; and the piebald white, with both the pure forms and many combinations evident in the modern domestic population.

The population genetics of each of the major coat-colour genes have been studied in some detail. The perpetuation of many highly distinctive coat genotypes in populations of domestic animals appears to be a product of man's preference for novelty; in wild animals a new mutation is often lost by chance, through genetic drift, or, if highly advantageous, displaces its original form and becomes fixed in the population. In the domestic cat, the effects of both artificial and natural selection can be seen. For example, the blotched tabby allele seems gradually to be displacing the wild-type striped tabby allele, even in feral populations, although the advantage that it confers has not been identified. This allele probably originated in Britain, sometime before the 17th century, and has been spread by colonization, particularly to Australia and North America (Todd, 1977). By contrast, the dominant white phenotype, although often preferred by people, confers disadvantages such as deafness and an increased susceptibility to skin cancers, which presumably act to suppress its frequency in natural populations.

In an urban population, Clark (1975) found that human preference had little effect on the selection of coat colours. While pet populations generally had lighter coats than did strays (at up to six separate genetic loci), the majority of the pets were neutered, and therefore the effect of human preference for colour was unlikely to be transmitted to the next generation. However, the global distribution of particular genes seems to be highly influenced by human activities, and particularly the long-standing custom of carrying cats on cargo boats. It appears that cats selected for this purpose tended to be of unusual types that might have disappeared had they remained part of a larger population. The sex-linked orange allele seems to have been affected in this way; it probably first became established in Asia Minor, and was then transported through the trading routes of the Mediterranean and, via the Seine and Rhone valleys, to London. However, existing large cat populations resulted in its only achieving a low frequency in these latter areas. Higher frequencies in Scotland and in North Africa appear to be a result of human preference for orange and/or tortoiseshell in those areas.

Domestication

Juliet Clutton-Brock (1987) describes the cat as 'an exploiting captive' and a 'carnivore that enjoys the company of man'. The cat is neither a man-made species like the dog, nor simply an animal made captive for utilitarian purposes, like the elephant. The history of its domestication is therefore unusual, and is also comparatively recent. Moreover, because there has been so little change in form from *lybica*, particularly in the skeleton, the archaeological evidence is often not conclusive.

The initial association between mankind and the cat was almost certainly commensal. The earliest grain stores, built at various locations in the Fertile Crescent some 12,500–11,500 years ago (YA), attracted rodents of various kinds, as well as leading to the evolution of the commensal mouse *Mus musculus*. Such a concentration of prey would have attracted small carnivores such as wildcats, some of which would have been accidentally pre-adapted to living in proximity with man and became tame. Some of these captive cats appear to have been deliberately transported to new centres of civilization, as evidenced by a cat burial in Cyprus (an island without a resident wildcat population) some 9500 YA (Vigne *et al.*, 2004). However, the archaeological record does not provide evidence for domestication of the cat progressing any further than taming for a further 4000–5000 years.

Domestication proper had probably started in Egypt by about 4000 YA, since remains of *F. silvestris* are found in Egyptian tombs from that period. Egyptian paintings and sculptures dating from about 3600 YA onwards give conclusive evidence of domestic status, as cats are depicted sharing many of man's activities, such as eating and hunting (Serpell, 2000). Theories of the initial function of domestication are divided between the utilitarian and the cosmetic. The economy of Egypt was, at that time, largely based on grain, and so the ability of cats to control outbreaks of granivorous rodents might have instigated encouragement from the populace, perhaps in the form of supplementary feeding and the provision of protected nest sites. According to this theory, North African wildcats initially became commensal, and then started the process of domestication themselves. It is also possible that wildcats were locally rare or absent, and that tame cats were deliberately brought into Egypt, where they were able to evolve into domestic cats without risk of hybridization with wild-type animals. Whatever the initial route, from 3500 YA cats came to play a part in the Egyptian pantheistic religions, culminating in the elevation of the cat-goddess Bastet to the national deity about 2950 YA. The sacred status of cats led to their being mummified in vast numbers, leading to the bizarre use of many tons of these corpses as fertilizer, following their excavation in the early part of the 20th century.

The spread of the domestic cat from Egypt appears to have taken place rather slowly, and indeed their religious status appears to have hindered their export as pets or as controllers of rats and mice. The Romans appear to have favoured the much less tractable weasel and polecat for the latter purpose, until about 1600 YA, when they began to prefer the cat. Domestic cats had reached India by about 2200 YA, and thereafter spread to the Far East. This eastward spread gave rise, probably via genetic drift, to a genetically distinct type of cat, now distributed throughout the Far East, from which the breeds characteristic of that area were subsequently derived. A second distinct population arose in Northern Europe, possibly brought there by Vikings 1200–1100 YA, and was subsequently transported to countries colonized by England, Portugal and other seafaring nations. A third genetically distinctive population has been identified in Kenya, and others may await discovery. The DNA of most cats in Italy, the Near East and North Africa still resembles that of the original domesticated population.

In modern times the distribution of the domestic cat has become global, but with marked differences between countries (Turner and Bateson, 2000). Cats are evidently most popular in the Antipodes, North America and Western Europe, although there are some cat-loving countries outside these areas, notably Algeria, Israel and Indonesia. The extent to which cats are 'owned' varies considerably from one country to another; for example, many cities contain large populations of strays and ferals. In the USA there has been a recent trend for cats to become more popular as pets: in 1983 cats, at 52.2 million, were owned by 28.4% of households; by 2007 they had overtaken dogs in numbers (81.7 million compared with 72.1 million) if not in the number of households (32.4% compared with 37.2%). In the UK, one recent estimate places the pet population at 10.3 million cats and 10.5 million dogs (Murray *et al.*, 2010). Increasing urbanization appears to favour cats over dogs, giving substance to the prediction that the cat is the pet of the future.

In behavioural terms, domestication has probably had less effect on the cat than on any other domestic mammal. The changes that have taken place seem to be of three kinds: (i) reduction in brain size, often correlated in other domesticated animals with a reduced sensitivity to uncongenial stimuli; (ii) modification of the hormone balance, mainly by reduction in size of the adrenals; and (iii) neoteny, the persistence of some juvenile behaviour characters in the adult. Certain of these changes may have become necessary as the population density of cats increased, as they moved from being territorial into agricultural communities and then into towns. However, the skulls of cats placed in Egyptian tombs – at a time when the cat was thought to have been at least partially domesticated – indicate that at that stage the brain was at least as large as that of modern *lybica*, and distinctly larger than that of modern *catus*. Neoteny is much less apparent in cats than in dogs, although such behaviour patterns as purring and treading with the forepaws, both characteristic of kittens, may be more commonly expressed in adult domestic cats than in their wild counterparts, although it is unclear to what extent this is innate and to what extent it is due to incidental reinforcement by attention from owners.

Association with man has little altered the cat's wild behaviour patterns. For example, domestic cats retain a fully functional repertoire of predatory behaviour, and also courtship behaviour and mate selection. Patterns of predatory behaviour are practised from an early age by kittens, and are used to considerable effect by adults (in most breeds of dog the predatory sequence is incomplete). This may be due to the cat's unusually stringent nutritional requirements, which until their elucidation in the latter part of the 20th century were difficult to satisfy except by hunting; critically, female cats have a dietary requirement for prostaglandin precursors, without which they are unlikely to breed successfully (Bradshaw, 2006).

Man's incomplete control of the sexual behaviour of cats (apart from pedigree animals) may help to perpetuate the maintenance of 'wild type' characteristics. Increasingly, mankind attempts to control the breeding of cats through neutering, rather than by the traditional disposal of unwanted kittens (depending on the area, 60–98% of pet cats in the UK are neutered; Fig. 1.2). However, those that are not neutered are usually allowed to select their own mates, perpetuating the mating system (Table 1.2). The effects of this can be seen in differences in the frequency of coat colours between areas where cats are allowed to breed freely and areas where most individuals are neutered; there may also be parallel effects on temperament and behaviour, although these have not been documented. Although breeding is much more tightly controlled in the 'fancy' breeds, where mates are selected on the basis of the predicted outcome of particular combinations of morphological characters, chance matings can result

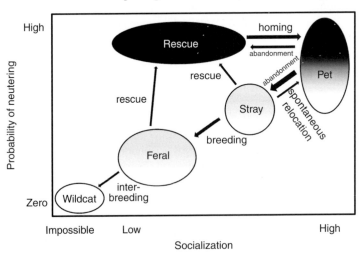

Fig. 1.2. Populations of *Felis catus* and *silvestris* in the UK, indicating typical extents of socialization (to people) and probability that an individual will be neutered, and the main processes leading to movement of cats or their offspring between one population and another. Interbreeding with Scottish wildcats, while commonplace where the two races co-occur, appears to result in hybrids that suit neither niche, tending to maintain the genetic and phenotypic integrity of each.

in the transfer of 'fancy' genes into the general population. Especially in areas where a high proportion of pet cats are neutered, stray and feral cats may make a substantial contribution to each generation of domestic pets, both through matings between feral males with owned females and the 'rescue' of kittens born to stray and feral females (Bradshaw *et al.*, 1999; Table 1.3).

In contrast to the situation in most domesticated animals, including the dog, specific breeds make up a minority of the cat population, as little as 6–8% in both the UK and the USA (but as high as 23% in urban Australia). There has been little direct study of the behavioural characteristics of even the commonest

Table 1.2. Populations of *Felis silvestris catus*, defined according to their degree of dependence upon humans. Feeding by people may be specific (i.e. provided on an individual basis), general (i.e. provided at a location but not specifically to any one individual), accidental (e.g. rubbish dumps) or none (i.e. a cat that subsists mainly or entirely by hunting). Shelter may be provided specifically (i.e. owned cats that spend much of their time in and near their owner's dwelling), accidentally (e.g. farm buildings) or not at all (cats living at some distance from human habitation, occupying, e.g. unused burrows of other species). Socialized cats display affiliative behaviour towards at least some people.

Population	Provision of food	Shelter	Socialized
Pedigree	Specific	Specific	Yes
Pet	Specific	Specific	Yes
Stray/semi-feral	General	Accidental	Yes
Feral	Accidental/general	Accidental	No
Pseudo-wild	None/accidental	None	No

Table 1.3. Probable constraints on reproductive success in the populations of *catus* defined in Table 1.2. Breeding may be either: (i) controlled (i.e. mate choice and frequency of breeding determined largely by the human owner); (ii) suppressed by neutering but without control of mate choice; or (iii) uncontrolled. See Bradshaw *et al.* (1999) for further details.

Population	Breeding	Mate chosen by	Reproductive success
Pedigree	Controlled	Human	Appearance, temperament
Pet	Suppressed	Cat	Avoiding neutering
Stray/semi-feral	Uncontrolled/ suppressed	Cat	Resistance to disease, nutrition, avoiding transfer to pet status
Feral	Uncontrolled (locally suppressed)	Cat	Resistance to disease, nutrition
Pseudo-wild	Uncontrolled	Cat	Hunting ability, resistance to disease

breeds, but given that purebred cats, unlike dogs, have been selected primarily for appearance rather than behaviour, major differences would not be expected. The experience of cat-show judges, breeders and veterinary surgeons has been used to produce overall descriptions (Hart, 1979; Takeuchi and Mori, 2009), although it is generally agreed that differences between individual cats can be as great as differences between breeds. Persians tend to be less active than other breeds; their long coats may make sitting on a lap for any length of time uncomfortable, possibly the reason why they seem to prefer to be stroked when sitting on the floor or on a chair. Abyssinians, which are short-haired, also tend not to be lap-cats. The extent to which such behavioural traits are inherited has not been determined, but they could be explained simply by changes in certain stimulus thresholds. An extreme example of this can be found in the poor maternal behaviour of many blue-eyed white cats. The combination of genes that brings about these colours also often results in deafness (Geigy *et al.*, 2007). Such queen cats are unable to hear the cries of their kittens, and are therefore less responsive to them. A further behavioural impairment is found in some Siamese cats, where the eyes and brain are in some way wrongly connected, resulting in impaired vision and, in some affected individuals, a compensatory squint.

Those with the slender oriental body type, such as the Siamese and Oriental Shorthairs, do show distinct behavioural differences from breeds with the stockier 'cobby' or occidental body type, such as the British Shorthair and the Persian. The oriental body type was first reliably described in Siam in the 18th century, but may well have originated elsewhere, at an earlier date. The average modern example of this type is, by comparison with typical cross-bred cobby cats, noisy, active, highly responsive to external stimuli and more trainable, suggesting a higher capacity to learn and/or a stronger social bond to the owner. Many Siamese cats actively seek out affection from people, although some authorities attribute this to a desire for warmth because of their thin coats. The Burmese types usually display Siamese characteristics, but to a lesser degree, and in particular are less vocal. There does appear to be a discontinuity between the oriental and occidental breeds, as reflected in their behaviour, which lends support to the DNA evidence that they are derived from different races of *F. silvestris*.

Relationship with Man

Cats have a unique degree of flexibility in their dependence on man. In the UK, something between one and two million cats are strays or ferals, relying on hunting, scavenging for waste food and handouts from cat-lovers, in varying proportions. It is unlikely that this population could be sustained at any substantial level without man's support; apart from the supply of food, in a temperate climate the survival of kittens is very poor except in the warm, dry conditions provided by human habitations. In warmer countries there is often less of a distinction between the house cat, the stray, the partially reliant and the truly independent feral, and in some areas, such as the Middle East, there appears to be little distinction between feral individuals of the domestic cat and wild *F. lybica*, raising the possibility that extensive interbreeding with the wild ancestor has occurred. This flexibility mainly stems from the cat's ability to adjust its social repertoire based on its experiences during early life (see Chapter 4, this volume).

During its co-evolution with man, the cat has retained this flexibility in order to survive the changes in the relationship that have taken place over the centuries, veering from worship to revulsion. In contrast with the deification of the domestic cat by the Egyptians, Christianity adopted the cat as an almost universal scapegoat, particularly from the 13th century onwards. Cats, particularly black ones, were regarded as a product of the devil. Their partly nocturnal habits, and the blood-curdling cries they make during fights, may have helped to single them out in this regard (Serpell, 2000). Similar associations with evil are common in Oriental folklore. This attitude persisted, at least in English-speaking countries, until the beginning of the 20th century, and there is still a substantial proportion of the public that expresses a dislike of cats. The reasons for this are obscure, but may be due to a combination of the old superstitions with distrust stemming from personal experience of the ambivalent relationship between cat and man. It may also relate to the apparent perceived 'independence' and lack of obvious facial expression in cats, as compared with the more overtly dependent and visually expressive characteristics of the dog, an animal derived from a much more social ancestor.

Leaving such prejudices aside, the relationship between cat and man can be described as mutually beneficial (Serpell and Paul, 2011; but see also Archer, 2011). The cat gains shelter and a reliable supply of food, both essential for the raising of healthy kittens. The owner may assist the cat in defending the core area of its home range, by discouraging other cats from stealing food provided for their own cat. Some of the cat's psychological needs may also be met through social interactions with the owner, such as play and displays of affection. The owner may benefit practically, through the control of vermin; in the UK, the majority of farmers keep several cats whose primary function is rodent control (Macdonald *et al.*, 1987). However, this role is one of decreasing importance for the majority of cats; indeed many modern owners are distressed by their cat's successes in the hunt. It is as companions that many people keep cats, satisfying their urge to nurture or, in a family context, giving their children intimate contact with a living being. In general, the relationship is not as strong as that between man and dog, in terms of both the amount of time put into the relationship and the degree of emotional investment, although this may be changing with the increased popularity of the cat as a pet. Cat owners value many attributes of their pets, including their cleanliness, reliability in returning home, lack of aggression towards people and other less easily described qualities such as 'femininity' and 'independence'.

General Biology

Skeleton and muscles

Many of the ways in which the cat differs from the general mammalian form are related to its carnivorous lifestyle. Unusually among domestic animals, its skeleton and musculature have been scarcely modified from the wild *F. silvestris*. While the skeleton adopts the general mammalian pattern, there is clear evidence of modification to permit efficient hunting. The most extreme example is the use of the front legs, which are made highly mobile by almost complete reduction of the collar-bone or clavicle, which is replaced by powerful muscles. This allows considerable fluidity of movement during normal prey-catching, which uses the claws of the front feet, and for balancing. Further flexibility of movement is achieved by highly mobile joints between the vertebrae. The tail is well innervated and acts as an effective counterweight, for example when the cat is attempting to walk along the top of a fence. Other adaptations that aid balance include specific changes in gait (Gálvez-López *et al.*, 2011), although the domestic cat is not as well adapted to tree climbing as some other small felids, notably the margay. The hind limbs are more normally jointed and are specialized for power, used particularly in jumping rather than running. The muscles of these legs tire quickly, and the fastest gait of the domestic cat is considerably less efficient than the corresponding gait in the cheetah *Acinonyx jubatus*, the only felid that specializes in running down its prey.

Studies of locomotion (summarized in Ewer, 1973) have indicated that the front and hind limbs play very different roles, particularly while the cat is walking. Most of the propulsive effort comes from the hind legs; at the moment when the front legs hit the ground they are sloping forwards, and therefore act momentarily as brakes. In the later part of each stride, the front legs do provide some forward propulsion, but this only cancels out the initial braking effect. The main function of the forelimbs in walking is therefore to take the weight of the relatively heavy front half of the body, while the rear limbs provide the net power. The 'fluid' character of the cat's walk is achieved by the form of synchronization between front and rear legs: as each front leg touches the ground and exerts its maximum braking effect, the hind leg on the same side is exerting its maximum forward effort. In the fastest gait, the gallop, the braking effect is reduced in a different way; the legs are already moving backwards when they meet the ground, and also the spine flexes at this moment, allowing the rear legs to continue their forward progress unchecked. Three variations on the gallop have been detected in cats. In the transverse and rotatory gallops the limbs never strike the ground simultaneously. They are distinguished by the type of synchronization between front and hind limbs, which strike in the same order (e.g. both right–left) in the transverse gallop, and in the opposite order (e.g. hind, right–left; fore, left–right) in the rotatory gallop. The third type, the half-bound, differs in that the hind limbs are touched down together, but the forelimbs are not. Some animals, but apparently not the cat, use the full bound gallop, in which the hind limbs touch down together, followed by both forelimbs together. In all three cases there is one flight phase per cycle in which all four legs are off the ground. This follows the last lifting of a foreleg, and ends when the first (or both) hind leg(s) strike the ground. Any one of the three gallops may be used at velocities between 2 and 6 m/s, the initial choice seeming to depend on individual preference, although the rotatory gallop may be used the most. Increases in speed are achieved by lengthening the stride and slightly increasing the flight phase. During extended galloping, cats can switch smoothly from one type to another, and also alternate which limb is leading and which trailing, by increasing or decreasing the time taken to lift each limb from the ground (Wetzel *et al.*, 1977). It has been suggested that by switching between the types of gallop, fatigue in individual limbs is minimized.

The skull is notable mainly for its large eye sockets, characteristic of a visual predator, and modifications to the teeth. There are only 30 of these, fewer than in many other carnivores (the minimum number in the Felidae is 26), and they are almost all adapted to meat-eating (Fig. 1.3), with the exception of the incisors at the front of the mouth, which are very small and are used mainly in grooming. The long, laterally compressed canines are used in holding food and specifically for dislocating the vertebrae of prey. These teeth are equipped with abundant mechanoreceptors, which may be used to sense the precise place that the killing bite should be delivered. Cats are unable to chew in the way that herbivores do, since their last upper premolars and lower molars, the carnassial teeth, act like shears to cut meat into swallowable pieces. The masseter muscle provides the power for this slicing action, performed when the mouth is almost closed. Considerable power is also required for the capture of prey, when the canines are being used, but under these circumstances the mouth is fully open, the masseter is not particularly efficient and a different group of muscles (the anterior temporalis and the zygomatico-mandibularis) provide most of the power (Fig. 1.4).

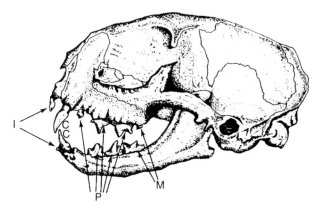

Fig. 1.3. Side (lateral) view of the skull of the cat, showing the arrangement of the teeth. I, incisors (3 upper and 3 lower on each side); C, canines (1 upper, 1 lower); P, premolars (3 upper, 2 lower); M, molars (1 upper, 1 lower).

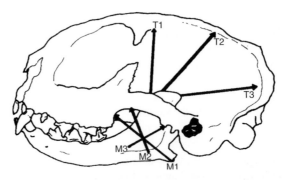

Fig. 1.4. Diagrammatic view of the skull, depicting the main jaw muscles as arrows. T1, T2 and T3 are parts of the temporalis muscles; M1, M2 and M3 indicate areas of the masseter muscle.

The way that cats ingest water has only recently been elucidated (Reis *et al.*, 2010). Unlike dogs, which curl the tip of the tongue to form a spoon that they use to scoop water into their mouth, cats pull a column of water into their mouth using the curled underside of the tongue. First, the tongue is pressed on to, but not into, the surface of the water. Then, as the tongue is retracted back into the mouth, pulling a column of water with it, the lips are closed, trapping the top of the column of water in the mouth. The tongue is then extended down again into the water, and the process repeated 3–4 times per second. The mouth thus gradually fills with water, which is swallowed every 3–17 cycles.

Skin and coat

The skin is loose, increasing the possibility that wounds incurred in fights with other cats or large prey will be superficial. The wild-type coat consists of long, coarse primary ('guard') hairs, growing singly from follicles, and a variety of shorter secondary hairs growing in groups. Selective breeding has resulted in many modifications, including long guard hairs (Angora), long primary and secondary hairs (Persian) and the absence of the guard hairs (Cornish Rex). Muscles on the roots of the guard hairs allow the coat to be erected, either to increase the apparent size of the body in social conflicts or simply as additional insulation in cold weather. The heavy coat creates problems for thermoregulation in hot weather; eccrine sweat glands are to be found only on the soles of the feet and cats frequently use the evaporation of saliva to lose heat, either by panting or by smearing saliva on the coat. To conserve energy, panting occurs at the natural resonance frequency of the respiratory system, about 250 cycles per minute.

Grooming, whether directly by licking the fur or indirectly by brushing with the front paws, is ultimately carried out by abrasive hooked filiform papillae on the centre of the tongue. The claws, made from the structural protein keratin, are derived from the skin rather than the skeleton. Often referred to as retractile, they are more properly termed protractile, since their resting state is sheathed. Each is attached to the final toe-bone, and is unsheathed by tendons that pull on that bone to pivot the claw forwards and/or downwards (Perfiliev *et al.*, 1998). The paws are not used equally: many female cats prefer to use their right paw, whereas many males show a bias towards their left paw, the adaptive significance of which is still unknown (Wells and Millsopp, 2009).

Reproduction

Many details of reproduction in the cat conform to the standard mammalian pattern, the main difference being the trigger for ovulation. In common with other felids and some other mammals such as the rabbit, the cat is an induced ovulator. Stimulation of the vagina during mating results in a surge of luteinizing hormone from the pituitary, whereupon the mature follicles burst, releasing their eggs. It has been claimed that this is an adaptation to infrequent contact between the sexes, allowing synchronization of ovulation and fertilization, but this is an unlikely explanation for the same phenomenon in the rabbit, a highly social species, and the true value of this system may be found in the mating system of the cat (see Chapter 8, this volume).

Conclusion

The domestic cat still carries most of the morphological characteristics that were evolved by *F. s. lybica* as a hunter of small mammals in North Africa, and has undoubtedly retained many of its behavioural characteristics as well, the notable exception being the capacity to form social attachments to humans. Its spread to temperate climates has depended on shelter provided, deliberately or accidentally, by man, and this has been particularly important for ensuring the survival of kittens, which are still susceptible to the effects of cold and damp. Domestication of the cat has been a long and drawn-out process, and is still not entirely complete, given that the majority of kittens born are from uncontrolled matings. For the first 6000 years or so, the relationship was largely commensal, with individual wildcats foraging on concentrations of rodent pests: some of these animals were evidently tamed. The next stage on the route to domestication, deliberate breeding within an anthropogenic environment, appears to have been pioneered by the Egyptians of the Middle Kingdom. The selection of *F. silvestris* for domestication appears to have been largely accidental, since many other species of small Felidae are readily tamed; domestication, once it had been achieved, resulted in a much more tractable animal that was able to displace its merely tame rivals as it was spread around the globe via man's activities. For roughly 4000 years its genome was essentially little altered from that of its wild ancestor. However, recent hybridizations with other small cat species have led to the creation of novel types of domestic cats and, if this trend continues, the cat of the 21st century may become far more genetically diverse than that of the 19th.

Sensory Abilities

Introduction

A full appreciation of the way in which an animal reacts to its environment depends on our understanding of the stimuli that the animal can actually perceive. There is a tendency to assume that mammals see, hear, smell and feel in the same way that we do, and therefore live in the same subjective world that we do. There are obvious exceptions that are generally allowed for, such as the echo-locating abilities of bats, but because of their very familiarity it is easily overlooked that our domestic animals have different sensory abilities from our own. The domestic cat has been a favourite subject for investigations of the workings of all the major senses, with the exception of olfaction, so knowledge of the information that reaches the cat's brain is reasonably complete. However, there are fewer accounts relating the cat's behaviour and ecology to its sensory abilities. This chapter is an attempt to portray what information the cat can most easily glean from its surroundings, and the extent to which that information overlaps with what we would gather ourselves if placed in the same situations.

The Vestibular System

Our own sense of balance acts almost entirely at the subconscious level, and therefore we do not give it the same level of importance as sight, sound, touch or smell. However, one only has to look at the degree of control that the hunting cat brings to the movement of its body, and especially the movement of its

© J.W.S. Bradshaw, R.A. Casey and S.L. Brown 2012. *The Behaviour of the Domestic Cat*, 2nd Edn (J.W.S. Bradshaw *et al.*)

head, to realize that balance has a great part to play in the success of the cat as a predator. The functioning of the principal organ of balance, the vestibular system, is now quite well understood (see Wilson and Melville Jones, 1979). A brief description of this organ follows, together with the special features that are known for the cat.

The vestibular system forms part of the inner ear, and consists of fluid-filled tubes of two types (Fig. 2.1). The three semicircular canals detect angular movement of the head in all three dimensions, although the canals are not precisely arranged at right angles to one another, nor are the corresponding pairs of canals on each side of the head all parallel to one another. The canals work on the principle that the fluid tends not to move when the head makes sudden changes of angle, and this relative motion, detected on the walls of the canals, provides sensory information about the change. The other type, consisting of the utricular and saccular otolith organs, primarily detects both linear motion and gravity. A ciliated sensory epithelium is covered by a layer of crystalline deposits of calcite, known as otoconia, which gets 'left behind' as the rest of the head accelerates; this deforms the bunches of cilia, which provide the sensory input.

In the cat, the alignment of the three semicircular canals is much closer to being at mutual right angles to one another (orthogonal) than in other mammals, for example man or the guinea pig. Presumably this makes the integration of information coming from the three canals simpler; if the canals are not at right angles then any motion will result in signals from at least two canals.

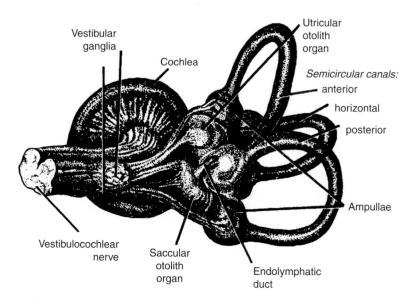

Fig. 2.1. The inner ear, showing the cochlea, which is concerned with hearing, and the semicircular canals and otolith organs, which are concerned with balance. Both are connected to a single sensory nerve, the vestibulocochlear, which connects to the medulla in the brain (not shown) (redrawn from Crouch, 1969).

Furthermore, the horizontal canal is precisely aligned with the normal position in which the cat carries its head. The utricular otolith organ is also 'tuned' to measure gravitational deviations from the normal head carriage most accurately. The role of the saccular otolith organ is less clear, but it does not seem to be particularly sensitive to gravity, so it may be specialized as a detector of movement.

The very precise carriage of the head in the cat, also seen in many other carnivores and notably in birds of prey, is largely a result of integration of information coming from the vestibular organ and translation of that information into precisely defined movements. While the more complex aspects of this behaviour are under the direct control of the brain, much of the output from the vestibular organ results in reflex movements of the neck and body muscle systems, and of the eyes. These reflexes, and their integration, are described in the next chapter.

Cutaneous Sensory Mechanisms

In common with the other senses, the sensory receptors on the skin of the cat are similar in structure to those found in other non-primate mammals. Their neurophysiology is quite well understood, as are the mechanisms whereby the information they produce is transmitted through the spinal cord and the cerebral cortex (Iggo, 1982). The cat's perceptual world cannot differ very much from our own, as far as touch is concerned, but a few special features can be discussed as part of a general description of the types of receptors that are present, and the kind of information that they produce.

There are as many as 15 different types of cutaneous receptor, or afferent unit, in the cat. They can be broadly divided into three categories: (i) mechanoreceptors, sensitive to touch and pressure; (ii) thermoreceptors, sensitive to temperature; and (iii) nociceptors, which in man produce subjective sensations of pain. Each group can be distinguished according to the stimuli that cause them to discharge. In most cases it is also possible to identify these neurophysiological categories with morphological characteristics, such as structures associated with nerve endings, and diameters and state of myelination of the afferent fibres. From a behavioural point of view, the most important aspect is the quality of stimulus that produces the response, for it is the sum of these qualities that defines the limits of the perceptual world of the cat.

The mechanoreceptors can be divided into two broad types. One responds mainly to movements of the skin or hairs with which they are associated, but not to sustained displacements. The other responds to both movements and displacements, for example sustained pressure as is felt on the pads of the feet. The former are known as rapidly adapting (RA) units, the latter as slowly adapting (SA). The resting rate of discharge from the SA units is proportional to the amount of indentation of the skin, or the extent of displacement of a hair or group of hairs. Type I SA units are often grouped in domed 'touch corpuscles', and sites where they terminate are totally enclosed in specialized structures called Merkel cells (Iggo, 1966). These units are particularly sensitive to stroking of the skin. The other common type of SA unit, SA II, is more widely distributed and terminates in Ruffini endings. These are more sensitive to stretching of the skin. A third type of SA unit, called the C-mechanoreceptor, signals lingering mechanical stimulations (those that stay in contact with the skin for 200 ms or more); in the cat, these receptors are more common on the hind quarters than the forequarters, but their function is not yet fully understood.

The RA units respond while displacements of the skin or hair are taking place, but become silent if the displacement is held steady. The hair follicles are innervated by their myelinated afferent fibres, which end in a ring around the sheath of the hair root. There are at least three types of hair follicle unit, each associated with a different type of hair (Table 2.1). The most sensitive are type D, which can be excited by movements of the down hairs alone; the receptive field of each unit can be as much as 2 cm^2, so each unit cannot individually give very precise spatial information. Type G cells are each associated with ten or more guard hairs, and each guard hair can be connected to several different units. The third classification, type T, is not as common in the cat as it is in the rabbit (Table 2.1); each unit is connected to between three and ten of the longest guard hairs, called tylotrich follicles. Found in both hairy and hairless skin are the Pacinian corpuscles, which are encapsulated receptors in the deeper skin layers. These are particularly sensitive to vibrations between 200 and 400 Hz, but in common with other RA receptors do not respond to steady-state displacements. This accounts for the cat's ability to detect vibrations through the pads of its feet even while those pads are being distorted by the weight of its body.

Different areas of the skin contain different numbers and proportions of mechanoreceptors. In the cat, the nose and the pads of the front feet are particularly well supplied, reflecting the use of the front paws in hunting and

Table 2.1. Characteristics of mechanoreceptors abundant in the hairy skin of the cat and the rabbit (saphenous nerve) (from Iggo, 1966).

Type of unit	Receptive field	Number of units	
		Cat	Rabbit
Rapidly adapting hair follicle afferent units			
Type T	3–10 longest guard hairs	13	41
Type G	>10 guard hairs	217	46
Type D	Guard and down hairs, large field (up to 2 cm^2)	89	91
Slowly adapting units			
Type I (epidermal)	Touch corpuscle, 1–5 per axon	113	77
Type II (dermal)	Spot-like, excited by stretching	39	9

manipulating food, although the paws of more manually dextrous carnivores, such as the raccoon, are even more highly innervated. Between the foot and toe pads of the cat are the highest densities of hair receptors anywhere on the body, and the glabrous skin is densely packed with several types of receptor (Burgess and Perl, 1973), interconnected in such a way as to produce tremendous sensitivity to speed and direction of movement of a stimulus across the pad. The system does not, however, discriminate between different textures presented statically, and it is thought that the cat gains information about surfaces by making sequential comparisons as it moves its paw over objects (Gray, 1966). On the nose, the glabrous rhinarium is equipped with groups of receptors in rete pegs, consisting of slow- and fast-adapting mechanoreceptors, and also temperature receptors (Abrahams *et al.*, 1987); the latter may be useful in detecting wind direction and thereby help to locate the source of windborne odours.

In addition to all this innervation of the pads and feet, there are highly sensitive and specialized SA cells in the soft tissue at the bases of the claws. These produce signals about the degree of extension and sideways displacement, which are separate for each of the claws (Gordon and Jukes, 1964). Indeed, it is almost possible to think of the feet as sense organs in their own right, and their degree of sensitivity may explain why many cats appear to dislike having their paws touched.

High concentrations of sensory units are also found around the whiskers (vibrissae), which are the familiar stiffened sensory hairs found on the head. Each of these functions as a mechanical transducer, conveying forces applied along its shaft to the follicle at the whisker base. These are well supplied with both RA and SA mechanoreceptors, so that information on the amplitude, direction and rate of displacement can all be sent to the central nervous system, and in addition the SA receptors are arranged so that information on static displacement is produced (guard hair receptors show little positional response). Similar arrangements of receptors are found at the bases of the shorter stiffened hairs around the lips. Other clusters of stiffened hairs, the carpal hairs, are also found on the wrists; in addition to SA and RA receptors, these are equipped with Pacinian corpuscles that can detect vibration, which the vibrissae cannot (Burgess and Perl, 1973).

Both the vibrissae and the carpal hairs provide sensory information about the position of the cat's head and legs in relation to nearby objects, and may also be sensitive to air currents. This information may be most useful in the dark, or when the cat is manoeuvring in a confined space, and may help to compensate for the cat's long-sightedness when objects are being manipulated close to the snout. The vibrissae are arranged in tufts, of which the largest are the mystacials or 'whiskers' (Fig. 2.2). These can be moved wholesale, both backwards for protection or forwards towards objects to be investigated (but are not as mobile as those of the rat). Unlike normal hairs, the whiskers are tapered, probably to improve the quality of information they can pick up about the texture of any surface they are touching (Williams and Kramer, 2010).

Fig. 2.2. Arrangement of the vibrissae on the head, drawn diagrammatically to show the position of each tuft. M, mystacials; S, superciliary; G1, G2, genals.

The sensory input from the whiskers to the brain is coordinated with visual information in the superior colliculus (Fig. 2.3). The large amount of nervous tissue devoted to processing and integrating information coming from the mystacials indicates their importance as sense organs, perhaps most importantly when the snout is being directed towards prey.

The other tufts are the superciliary, which act like extensions of the eyelashes in triggering the protective eyeblink reflex, and the two genal tufts on the cheeks, which are close to enlarged skin glands, and may act as scent-spreaders as well as tactile sense organs (Ewer, 1973). Many carnivores also have a tuft of vibrissae under the chin, the inter-ramal tuft, but this is absent from the cat family. Since cats do not tend to hunt with their noses to the ground, or dig with their noses, they may not require these vibrissae to the same extent as other carnivores. It is also possible that stiffened hairs in this area would interfere with the dispersion of scent from the submandibular glands (see Chapter 5, this volume).

The remaining types of cutaneous sensory units respond to temperature (thermoreceptors), and to severe mechanical and thermal stimulation (nociceptors). There are two basic types of thermoreceptor: warm receptors with non-myelinated afferent fibres; and cold receptors with myelinated fibres. Both respond to absolute temperatures, and to changes in temperature. The warm units are maximally stimulated by steady temperatures of 40–42°C, and also increase their rate of firing as the skin temperature increases. The cold units increase their firing rate as the temperature drops, and respond most powerfully at steady temperatures of 25–30°C.

Discharge of nociceptors is associated with sensations of pain in human subjects, but their presence in the skin of the cat does not necessarily mean that cats feel pain as we do. These receptors fulfil a protective function, necessary because both mechanoreceptors and thermoreceptors reach their maximum discharge rate (and can therefore give no further information) to

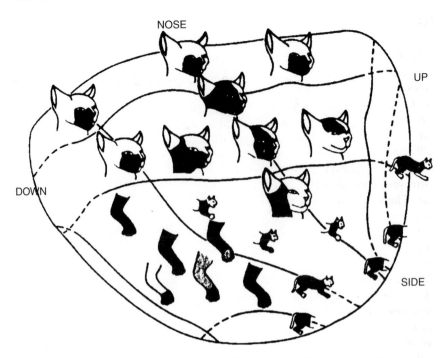

NOSE

UP

DOWN

SIDE

Fig. 2.3. A schema of the surface of the left superior colliculus, part of the mid-brain that processes and integrates sensory information. For each part of the surface, diagrams indicate from which area of the skin (shaded) tactile information is processed. The relative sizes of the diagrams indicate the amount of nervous tissue devoted to each area, and illustrate the importance of touch sensations coming from the face and the forepaw, and the relative unimportance of the body and the hind limbs. The contour lines indicate from which direction each part of the superior colliculus processes visual information. The horizontal plane, as seen by the eye, is indicated by the line running from NOSE (i.e. towards the nose) to SIDE (i.e. to the side), and the vertical plane by UP to DOWN. It can be seen that the visual and tactile information is approximately superimposed. For example, the part processing tactile information from the top of the head also processes visual information from the upper central part of the visual field; touch sensations from the forepaws are processed by the same part in which their visual image would normally be represented (down and out to the side of the field of view) (redrawn from Stein *et al.*, 1976).

stimuli that are intense, but not severe enough to be potentially damaging. Nociceptors consist of unencapsulated nerve endings of two types, a mechanical type that responds to squeezing and crushing of the skin, and a mechano-thermal type that also responds to extremes of heat and cold. Nociceptor responses are probably poorly localized, due to convergence of the neurons, but spatial information may be provided by the other cutaneous receptors that are inevitably discharged at the same moment.

Hearing

Vibrations in the ground may be primarily detected by the Pacinian corpuscles in the feet, and at the bases of the carpal hairs, but airborne vibrations are detected as sound by the hearing system. The latter broadly conforms to the general mammalian pattern, both in structure and function. The ears are commonly thought of as consisting of three units. The outer ear, consisting of the pinna, the concha and the canal, channels the sound, still consisting of airborne vibrations, to the middle ear (Fig. 2.4). This consists of the eardrum or tympanum and the ossicles, small bones that transform variations in air pressure into variations in fluid pressure in the inner ear. These vibrations are picked up by a variety of nerve cells, some tuned to particular frequencies and some to specific changes in frequency. The former permit the discrimination of two frequencies presented simultaneously (to take an example from music, the constituent notes of a chord). The latter can be tuned to species-specific sounds – for example, click-specific detectors are used by bats in echolocation. In man, both the outer and middle ear are structured so as to boost those frequencies that are particularly important in speech; the cat's pinna functions as a directional amplifier, boosting frequencies in the range 2–6 kHz (Martin and Webster, 1989), possibly an adaptation to the detection of species-specific calls. However, the bulk of the evidence suggests that many aspects of hearing in the cat have been shaped by the need to detect prey.

The detection of sounds

The audiogram of the cat, its ability to detect pure tones over a range of frequencies, is about 10.5 octaves, among the broadest recorded from any mammal (Heffner, 1998). For comparison, that of man is about 9.3 octaves. The structure of the cat's head, and the distance between its outer ears, has been compared with that of other mammals, and these comparisons predict that the hearing range has been extended at both the high-frequency and low-frequency ends; most mammals appear to have 'traded off' high frequency ability for low, or vice-versa (Heffner and Heffner, 1985). The cat-like carnivores (Feloidea) may have achieved this by the addition of a bony septum that divides the resonant space of the middle ear into two, forming the tympanic cavity and the bullar cavity (Huang *et al.*, 2000).

At low frequencies (>50 Hz), the cat's thresholds are broadly similar to those of man, a remarkable ability considering the much smaller head of the cat. The biological function of this ability has not yet been elucidated. In the middle frequencies (1–20 kHz), the cat is one of the most sensitive mammals; some studies have indicated thresholds down to between −20 and −25 dB SPL, compared with −5 dB for a young adult human, although other studies, such as that illustrated in Fig. 2.5, indicate lower sensitivity.

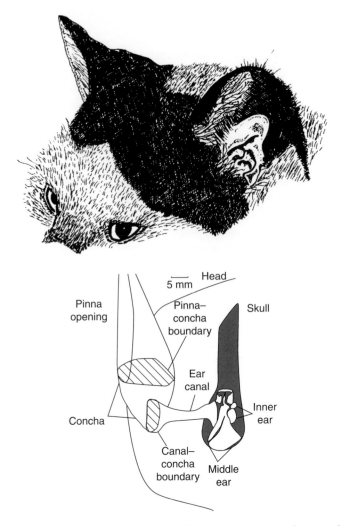

Fig. 2.4. The external ear (pinna) showing the complex corrugations on the inner surface that assist in the location of sources of sounds. The positions of the middle and inner ear are shown.

Reproducible differences can be detected in this region between the abilities of individual cats, which may explain some of the discrepancies. Some studies have detected a slight diminution in sensitivity, of unknown significance, at about 4 kHz. While man's thresholds decrease above 4 kHz, the lowest thresholds are at around 10 kHz for the cat, which can still detect sounds at 80 kHz if these are reasonably loud (the human equivalent would be louder than normal conversation but quieter than shouting). Because there is no abrupt cut-off in sensitivity at high frequencies, it is difficult to quote an exact figure for the upper limit, but it

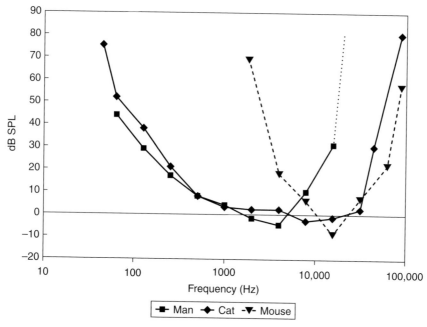

Fig. 2.5. Auditory thresholds for the detection of pure tones in man, cat and mouse. The thresholds are expressed as decibels of sound pressure level (SPL) – the lower the figure, the more sensitive the hearing at that frequency (high notes have high frequencies, and the region above the human range is normally referred to as ultrasound). Note that the hearing ranges of man and mouse hardly overlap at all, but that of the cat overlaps substantially with both (data from Fay, 1988).

is generally accepted that the useful limit for the cat is about 60 kHz. This ability is particularly remarkable because the efficiency of the cat's middle ear drops at frequencies above 10 kHz, due to mass limitations in the ossicles and to 'break up' of the eardrum into smaller vibrating portions. This high-frequency ability is presumably related to the detection of ultrasonic calls emitted by small rodents, and is therefore an adaptation to hunting by sound cues.

In the ability to discriminate between sounds of the same frequency but of different intensities, the cat is similar to other mammals but inferior to man. Man is also better at discriminating pairs of sounds of the same intensity but of different frequency, provided the frequency is below 5 kHz. Above this figure, cats outperform man although significant differences between the abilities of individual cats are evident. However, it should be noted that some of these interspecific differences may be due to the study procedures used. The comparisons just made rely on the results of conditioning experiments (cats) and verbal reports (human subjects). Neurophysiological measures from cochlear nerve fibres indicate that cats have the potential to be as discriminating as man at low frequencies, and to be considerably better at frequencies over 2 kHz. Cats also appear to be less sensitive than man in the detection of sounds of very short duration (Costalupes, 1983); man's abilities in this area have been linked to the complexity of our vocalizations.

Localizations of sound in space

All of the types of discrimination considered so far could be done as well with one ear as with two. Cats that are deaf in one ear do learn to locate sounds, often by making exaggerated scanning head movements to produce alterations in the intensity and quality of the sound at the single working ear. However, such cats never approach the abilities of those with complete hearing, showing that comparisons of the sound arriving at the two ears are critical to the pinpointing of the source of a sound, a skill that is critical to successful hunting in cover. The problem facing the cat can be broken down into the position of the sound source in the horizontal plane, the plane in which its ears are normally separated, and in the vertical plane, in which they are not.

Location of sound in the horizontal plane is achieved by comparison of the different signals arriving at the two ears, but the type of comparison that can be made is an inevitable function of the physics of the sound. For signals below about 4 kHz, for which the wavelength is less than the distance between the pinnae, time comparisons are made; the relatively small head also makes comparisons below 500 Hz difficult. At frequencies above 6 kHz, time differences become ambiguous and the cat relies on differences in intensity, produced by the masking effect of the head itself and the directional properties of the pinnae, such that more high-frequency sound reaches the nearest ear (Huang and May, 1996). There is a range of frequencies over which neither method is particularly accurate, resulting in a decline in angle-separating ability between 4 and 12 kHz, before the discrimination increases again between about 14 and 24 kHz. Of course, a hunting cat is unlikely to encounter pure tones, and the minimum angle that can be discriminated for noise signals that contain a broad range of frequencies is lower than for any pure tone.

The pinnae play a crucial role in the location of sound. In man, the corrugations add reflections to the spectrum of incoming sounds, giving information on the elevation of the sound source and whether it comes from in front or behind, for which comparisons of the signals from the two ears are ineffective. The cat gains an additional advantage from the mobility of its pinnae, which is under subcortical control and is accordingly extremely rapid (Tollin *et al.*, 2010). Most cats find localization of sound in the vertical plane difficult if the sound consists of a single frequency, but this is very accurate if the sound is of mixed frequency (Martin and Webster, 1987). This indicates that the cat, like man, uses spectral transformation cues to determine elevation. At frequencies above 12 kHz, the overall shape of the pinnae and the corrugations within them produce complex differences in the signals reaching each ear – for some combinations of elevation and frequency the signal is actually louder in the more distant ear (Martin and Webster, 1989). By comparing the intensities of different frequencies, the cat has sufficient information to localize the height of the source of the sound accurately. This ability must presumably be learned, as the shape of the pinna varies from one individual to another and changes as the cat grows.

Vision

The eye

Although the cat's eye conforms to the standard mammalian pattern, there are specializations that result in its visual world being very different from our own. Table 2.2 illustrates some of the differences between the structure of our own eye and that of the cat. Broadly speaking, the cat is much better adapted to vision at very low light intensities than we are. The maximum density of rods, the most sensitive visual receptors, is almost three times our own, while there are far fewer cones, the less sensitive colour-detecting visual receptors that we use in daylight. The centre of our own visual field corresponds on the retina to the fovea, an area in which rods are absent and cones are present at high density. Cats have no fovea; in their corresponding *area centralis* the density of cones is six times lower, and many rods are also present. The innervation of the eye is also differently organized; man's optic nerve contains over ten times more fibres than that of the cat, and there is an even greater discrepancy in the maximum density of ganglia in the retina itself. Overall, this results in more rods and cones being connected to each nerve, an adaptation towards greater sensitivity and therefore better night vision; this adaptation has its costs, as we shall see. The efficiency of vision

Table 2.2. A comparison between ocular and retinal measurements for cat and man (from Berkley, 1976).

Parameter	Cat	Man
Ocular		
Eye diameter (mm)	22	25
Total power of ocular optics (D)	78	60
Pupil		
Maximum diameter (mm)	14	8
Maximum area (mm^2)	160	51
Minimum diameter (mm)	<1 (width)	2
Interpupillary distance (mm)	36	62
Retina		
Maximum cone density ×10^3 (/mm^2)	26	146
Minimum intercone distance (µm)	6	3
Maximum rod density ×10^3 (/mm^2)	460	160
Maximum ganglion cell density ×10^3 (/mm^2)	3	147
Optic nerve and chiasma		
Total number of fibres ×10^3	85	1100

is still further enhanced by the tapetum, a layer of reflective cells immediately behind the retina, which reflects back any light that has not been absorbed by the visual pigments in the rods and cones at the first pass, giving a second chance for absorption before the light passes back out of the pupil again. It is this process that gives cats their 'eye-shine' when a strong light is shined at them when it is dark and the pupil is wide open. Tapeta are found in a variety of vertebrates; the type found in the cat, the *tapetum cellulosum*, is also found in some other terrestrial carnivores, seals and certain lower primates. The reflecting material consists of rodlets, 230 nm in diameter, arranged in regular arrays within cells in the choroid. The reflective properties seem to be due to the presence of riboflavin. In the cat, the tapetum increases the efficiency of the eye by about 40%.

Cats have large eyes in proportion to their body size, and the path of light from the pupil to the retina is shorter than in man, increasing efficiency further. Their pupils can be opened wider than can ours (Table 2.2), but in order to protect the sensitive retina, and to allow vision under bright conditions, they also have to be capable of closing to a smaller area than ours need to be. A simple circular arrangement of muscles around the iris appears to be inadequate to generate this extreme range of pupil size, and rather than contracting from a large circle to a small one the cat's pupil contracts to a fine slit, less than 1 mm wide. When the pupil is fully open, retinal illumination is of the same order as that of nocturnal creatures such as the badger and bat. Diurnal primates, such as man and the chimpanzee, have about five times less retinal illumination (Hughes, 1977).

The cat's slit pupil has also been proposed to be an adaptation to colour vision. The cat's pupil is large in order to gather as much light as possible: this incurs the disadvantage that if light of one colour is brought into focus, other colours (i.e. other wavelengths) are inevitably out of focus. To overcome this, the cat's lens is multifocal, such that the focal length changes from the centre of the pupil out towards its edge (Fig. 2.6). An iris that contracted to a pinpoint in bright light, as in the human eye, would allow only one colour at a time to be brought into focus, but contracting to a slit allows the full circumference of the lens to be used (Malmström and Kröger, 2006).

Cats are slower to transfer the focus of their eyes from near to distant objects, and vice versa, than are humans. The ability of an individual cat to transfer focus may also vary depending on early visual experience: outdoor cats tend to be slightly long-sighted, whereas cats raised indoors, where the furthest objects on which they can focus are only a few metres away, become short-sighted (Belkin *et al.*, 1977). When it is completely relaxed, for example during sleep, the cat's eye is long-sighted and has to be brought into focus on awakening. The normal degree of accommodation in the awake cat should produce a reasonably clear image of any object more than 2 m away, and the furthest degree of accommodation that has been observed, up to about 4 diopters, should produce clear vision at about 25 cm. The behavioural evidence suggests that cats cannot focus on objects closer than this (Elul and Marchiafava, 1964). Some stimuli elicit much more accommodation than others; a clump of feathers is far more effective than a mouse or a kitten at the same distance. Moving

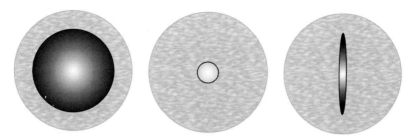

Fig. 2.6. The domestic cat has a multifocal optical system in which the centre of the lens focuses on a different colour (shown here as shades of grey) compared with the edge (left). A circular iris, when contracted in bright light, would allow only one colour to be in focus at a time (centre). The cat's slit pupil allows all zones of the lens to be used even when it is contracted (right) (redrawn from Malmström and Kröger, 2006).

stimuli are also more effective than stationary ones, suggesting that cats may only focus their eyes accurately if detailed information is required.

At least in adult cats that have been reared indoors, the lens is stiff. Therefore, distortion of the lens, the main focusing method in the human eye, is unlikely to be effective for the cat. When cats are looking at objects close to their face a bulge can be seen on the iris, suggesting that most of the accommodation to distance is achieved by the whole lens being moved backwards and forwards.

The eyes are placed on the head so as to point at about 8° off the centre line; the eyes of most mammals diverge much more than this, although felids tend to have small divergences (that of the lion is about 5°). This restricts vision to the area in front of the head (Fig. 2.7), an adaptation to a carnivorous lifestyle, since animals that are preyed upon need all-round vision for protection. The total visual field extends to about 200° in total, with binocular vision in the central 90–100°, figures not dissimilar to those for human vision. Within the region of binocular overlap, we can assume that the images are fused, so that cats have single vision. The eyes can be seen to make conjugate movements as if each is fixated on the same object, and identical changes in focus occur in both eyes even if only one eye can actually see the object that is being focused on. As objects are brought closer to the face, vergence movements can be observed, that is, the eyes are brought into a slight squint. Feral cats are reported to display more accurate vergence movements than cats reared indoors, so developmental factors may have a part to play in allowing binocular vision at close range (Hughes, 1972). The limit of vergence for many cats seems to be between 10 and 20 cm, over which range focus is probably also poor. This suggests that vision plays little part in directing the close-range manipulation of prey, for which tactile stimuli are probably much more important, accounting for the cat's sensitive vibrissae.

Despite their relatively large eyes, cats are able to make very rapid eye movements in response to moving objects. If an object moves very quickly, or

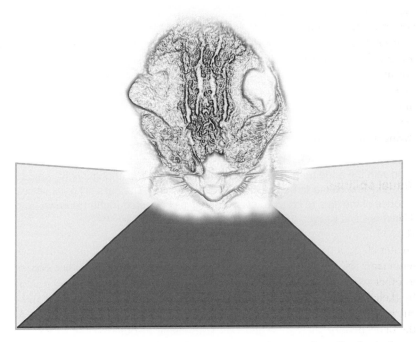

Fig. 2.7. The cat's field of view, including the binocular area (heavily shaded).

suddenly appears off the centre of the cat's line of sight, characteristic movements of the eye, known as saccades, are made. These consist of a lag phase of
about 0.25 s, during which the sensory information is translated into instruc-
tions for the ciliary muscles, a rapid phase of acceleration and then a longer
deceleration phase. While the saccade is occurring the position of the object is
continuously monitored, and if it moves again a second saccade can immediately follow the first to bring the eye back on to the target. That these second
saccades are usually successful is a tribute to the ability of the nervous system
in making relative calculations of the position of the object on the retina, and
its actual position in space. Horizontal and vertical saccades are brought about
by different pairs of extra-ocular muscles, and are controlled by different areas
of the brain stem. The vertical movements are the faster (velocities of up to
$250°/s$ have been recorded); the horizontal movements reach a maximum speed
of about $100–150°/s$ over a wide range of total angular displacements. If the
object that is followed moves diagonally, the corresponding saccade proceeds at
the pace of the slower component, i.e. the horizontal.

If the change in angle is small – about $8°$ or less – a saccade may occur,
but cats have an alternative slow eye movement, lasting over twice as long as a
saccade, that has not been observed in primates. The biological significance of
these alternatives is unknown. The slow eye movements are also distinct from

the smooth eye movements used to track slowly moving objects. Cats are much poorer at this than are primates, and tend to lag behind the object, catching up with small saccades at speeds as slow as 2°/s. If the whole background moves, as it inevitably appears to do when the head is rotated, the maximum speed of slow eye movements increases to about 8°/s, but only in the horizontal plane. The maximum is slower upwards, and much slower downwards; abrupt downward movements of the visual field occur during normal walking, so the latter may be necessary to prevent exaggerated eye movements being triggered by locomotion (Evinger and Fuchs, 1978).

Visual abilities

The absolute threshold for detection of light by the cat is about 3×10^{-7} cd/m^2, about 3–8 times better than man's. This difference is almost entirely due to the greater efficiency of the cat's eye, as described above. In both cat and man, about ten quanta of light, spread over the whole retina, are enough to cause a response; in other words, both systems are reaching the limits of sensitivity as determined by their biophysics. Cat and man are also similar in their sensitivities to different wavelengths at low light levels (i.e. when rods, and not cones, are active); the only small disparity is a small peak at 560 nm in the cat's spectrum, which corresponds to the frequency most efficiently reflected by the tapetum. However, our own vision is superior in several respects to that of the cat. For example, in moderate lighting cats are less able to discriminate which is the brighter of two lights, although at low light levels there is less disparity (Berkley, 1976).

At the onset of darkness, the human eye takes 30–40 min to become fully sensitive, due to the necessity for biochemical processes in the visual pigments. In the cat, such total adaptation can take up to 60 min if the pupil has been artificially dilated, but it has been suggested that cats may protect their dark adaptation by constricting their pupils when the light level increases suddenly. As light levels gradually increase the rods adapt, initially with changes in the way that signals from individual rods are combined and then, as light levels approach those at which the cones become functional, each rod becomes individually less sensitive (Tamura *et al.*, 1989). Of course, the timing of dark adaptation is unlikely to reduce the efficiency of vision at any time of day under natural light, because at dusk and dawn light levels change relatively slowly; it is our use of artificial light that has brought this inadequacy to our, and presumably our cats', attention.

Under conditions of moderate light intensity (2–75 cd/m^2), cats pick out far less detail than we do. Their threshold for distinguishing striped black-and-white patterns from uniform grey is about five stripes per degree of arc, while we can distinguish patterns that are six times finer. This can be attributed to at least three separate factors – the scattering of light by the tapetum, the smaller number of cones on the cat's retina and the greater numbers of rods connected to each retinal ganglion (Pasternak and Merigan, 1980).

There are various other standard psychophysical measures that can give an idea of the amount of detail that the cat's visual system can resolve. The critical fusion frequency (CFF) measures the point at which a flashing light is seen as constant illumination, a critical factor in our own perception of films and television. At low levels of illumination (0.03 cd/m^2), cats can just distinguish 60 cycles/s as flickering, better discrimination than our own, even when the lighting level is corrected for the cat's more efficient eye. This means that many cats are likely to see fluorescent tubes and televisions as giving off a flickering light. However, the cat's perception of slower flickering can be poorer than our own, particularly if, rather than the light being extinguished between bursts, it is merely made dimmer (Berkley, 1976).

Such temporal responses are presumably linked to the ability to detect movements and shapes. Considerable progress has been made in understanding the neurophysiology of the transformation of the simple on/off responses of the rods into the detection of particular shapes and patterns of movement. As the information moves along the visual pathway to the visual cortex, the neurons respond to increasingly sophisticated combinations of light and shade (Clifford and Ibbotson, 2003). In the visual cortex the majority of neurons are directionally selective, and many are also sensitive to the orientation of the stimulus and even for particular velocities of movement. Behavioural measures of the cat's abilities in this area have lagged behind, but several studies have indicated a sophisticated level of visual information processing. It is known, for example, that cats can discriminate small differences in the shapes of geometric figures such as triangles. They are also able to make relative size judgements; when trained to discriminate the larger of a pair of figures, the larger was still picked out even when the absolute size of both was altered. White-on-black shapes are seen as similar to the black-on-white version of the same shape (Berkley, 1976).

The detection of movement is an obvious necessity for hunting, but cats seem rather poor at detecting slow movements. The slowest angular speeds that they can be trained to detect are in the range of 1–3° arc/s, whereas we can detect speeds about ten times slower. The direction of movement, up, down, left, right or diagonally, has little effect on the threshold (Pasternak and Merigan, 1980). The finest distinctions of velocity can be made at speeds of 25–60°/s, which presumably correspond to the speeds at which prey are likely to move, since accurate estimation of angle and speed must be crucial to a successful pounce.

Cats appear to be able to join partly hidden outlines together, thought to be a component of the mechanism that allows the separation of 'figure' from 'ground' (Nieder, 2002). Testing this relies on methods based on illusions such as that shown in Fig. 2.8; in the upper set of figures, the eye constructs an illusory black square that can be made to appear to move down the background. If the white sectors are aligned in other ways, as in the lower set of figures, no such squares result. Training experiments have shown that cats can discriminate the figures that produce illusory squares from those that do not (Bravo et al., 1988).

Another way that we separate objects from their background is by discontinuities in texture. Cats can discriminate between the two images shown in Fig. 2.9, showing that they too can use texture to segregate images (Wilkinson, 1986). Such experiments suggest that the retinal image is segmented into biologically relevant features at a relatively low level of neural machinery, and that perception of such illusions does not involve cognitive processes. While no simple pattern-recognizer has been identified in the cat, akin to the frog's 'fly-detector' that directly elicits

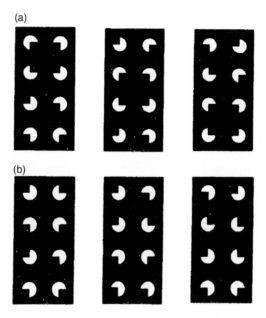

Fig. 2.8. (a) Viewed in sequence, the upper set of pictures appears to show a square that descends from the top set of discs to the bottom, while all the other discs appear to spin at random. (b) In the lower set, all eight discs appear to spin. Cats can discriminate between the visual illusion shown in (a) and the random pattern in (b), indicating that they do see subjective contours (redrawn from Bravo *et al.*, 1988).

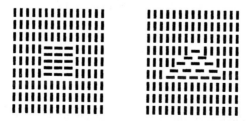

Fig. 2.9. Cats can discriminate between subjective shapes that are formed by changes in texture, such as the examples shown (Wilkinson, 1986).

feeding movements, there is now no doubt that the cat's detection of movement and pattern, and their interactions, is highly sophisticated.

Colour vision

At one time it was thought that cats saw primarily in monochrome, and although it is now known that they can distinguish between colours under daylight conditions, they seem to place little emphasis on colour when evaluating what they are looking at. There is clear neurophysiological evidence for two types of cone, one green-sensitive peaking at wavelength 560 nm and one blue-sensitive, at 455 nm. It is thus reasonably certain that cats are dichromatic, seeing two pure colours and their combinations, rather than the three pure colours that most humans can see. The lack of red-sensitive cones means not only that cats do not see the colour red, but also that red objects appear much darker to them than they do to us, compared with green or blue objects.

Under dim light the cones are inoperative and, since all the rods are maximally sensitive to light of wavelength 497 nm, cats will see in black and white, just as we do. At background illuminations in the range 3–30 cd/m^2 the cones take over and two peaks of sensitivity can be detected, one produced by the green- and the other by the blue-sensitive cones. The greater sensitivity is to the blue, and this cone also dominates both temporal and spatial resolution when sufficient light is available (Loop *et al.*, 1987).

Some idea of the unimportance of colour for cats can be gained from the number of wavelength-comparing ganglion cells that can be detected, about 16 times fewer than in primates. This has led to speculation that cats not only perceive fewer colours, but that these colours are much less saturated than those that we perceive.

It has proved very difficult indeed to train cats to respond to different colours of the same subjective brightness. Some investigators never succeeded, while others demonstrated weak effects after thousands of trials, or many months of continuous training. It is these results that have led to the view that cats are behaviourally colour blind, whatever the physiological evidence. The most likely explanation is, however, that colours provide poor training cues for cats, because of perceptual dominance. This means that cats are much more likely to associate other visual cues, such as pattern or brightness, with the rewards they obtain during training, and therefore take a very long time to realize that the appropriate cue is actually colour. However, Tritsch (1993) was able to train cats to discriminate reliably between blue and orange even when their relative brightness was varied by changing the composition of the light under which they were viewed.

Stereoscopic vision

The cat's visual cortex contains binocularly activated nerve cells that respond to retinal disparities (DeAngelis, 2000), so it is highly likely that cats see in three dimensions (3D). Behaviourally, cats can distinguish between a large and a small object of the same shape, even when the large object is so far away that its image on the retina is smaller than that of the small object (Gunter, 1951). One-eyed cats can make comparisons of depth, but while doing so they make exaggerated head movements that would provide a temporal version of binocular vision. They are also much less accurate at judging distance than normal cats. The best evidence is that cats, like us, use retinal disparity to judge depth; that is, they rely on the double images that are formed of objects both in front of and behind the plane of fixation. The value of this ability for hunting is, of course, considerable.

The visual system

The neurophysiology of vision is probably better understood in the cat than in any other non-primate mammal (e.g. Troy and Shou, 2002; Loffler, 2008), and yet the behavioural significance of this physiology is still imperfectly understood. At this point, knowledge of the neurophysiology is still in advance of the behaviour; for example, pathways for the parallel processing of information about spatial frequency, motion, colour and binocular disparity have been detected, but the resultant effects on behaviour are still poorly understood. However, we now understand a great deal about the visual world in which the cat lives, which is helpful for the interpretation of responses to biologically meaningful stimuli.

It should be remembered that the comparisons that have been made with the abilities of primates, particularly man, often rely on different methodologies for each species, and may therefore be unreliable. Difficulties in training cats to respond to particular stimuli may lead to false conclusions, as, for example, happened for colour vision. There has also been a tendency to examine aspects of vision in the cat that are familiar from our own visual experience; some of the cat's visual abilities may still be undetected, simply because they have no human counterpart.

Abnormal vision in the Siamese cat

Vision in Siamese cats differs from that in non-orientals in several ways, one of which is a reduced ability to detect flicker. In the retinas of all cats two types of ganglion cells can be detected: X-type, which respond to patterns of luminance; and Y-type, which respond to movements and also inhibit the output of the X-cells while the eyes are moving. Siamese cats have only about 14% of Y-cells, compared with 35–45% in cross-bred non-orientals (Loop and Frey, 1981). What these differences mean in terms of the subjective impressions of vision in Siamese cats is difficult to say precisely, but they may be less sensitive to flickering lights, while their vision may be temporarily impaired, possibly by blurring, when they are moving their attention from one object to another.

More serious may be their poor stereoscopic vision (Bacon *et al.*, 1999). Most of the nerve fibres from each eye cross over in the optic chiasma and innervate the contralateral side of the brain; in normal cats about 35% of the fibres stay on the same side, enabling comparisons to be made of the disparities between the images received by each eye. Few binocularly driven calls can be detected in most Siamese cats, and no behavioural evidence can be found to indicate that they have stereoscopic vision, although the acuity of each eye is apparently normal. Probably as a result, Siamese cats have a tendency to develop convergent squints.

The Chemical Senses

There are now acknowledged to be at least four distinct chemical senses in most mammals: taste, olfaction, the vomeronasal and the trigeminal. Two of these, taste and vomeronasal, have been specifically studied in cats. Since the sense of taste is intimately bound up with food selection and the detection of nutrients, it will be described in the chapter on feeding behaviour. Published accounts of the olfactory and trigeminal senses have tended to focus on other mammals (e.g. Firestein, 2001), and so the descriptions that follow necessarily assume that the cat is not very different from other mammals, particularly the domestic dog.

Olfaction

The sense of smell is very important to cats, as it is to the great majority of non-primate mammals. Some idea of this can be gained by the behavioural changes that occur when cats lose their sense of smell as a result of virus infection, particularly feline viral rhinotracheitis. Such cats frequently lose their appetite, change their toileting habits and do not indulge in courtship. The damage may well extend to the trigeminal and vomeronasal systems as well, so these changes do not indicate that they are triggered by exclusively olfactory stimuli. More direct evidence comes from the size of the olfactory epithelium, which can be 20 cm^2, compared with 2–4 cm^2 in man. This epithelium is supported on some of the scroll bones, the ethmoturbinals, while others, including the maxilloturbinals, serve to clean, warm and moisten the inspired air before it reaches the delicate olfactory receptors (Fig. 2.10). The maxilloturbinals, while moderately well developed in the cat, are even more extensive in more active carnivores such as the dog, where more rapid respiration during exercise will bring greater volumes of air into contact with the olfactory apparatus.

The detailed structure of the olfactory epithelium itself has been revealed by electron microscopy, and is considerably simpler than, for example, that of the retina (Fig. 2.11). The surface is protected from direct contact with the air by a layer of mucus secreted by the Bowman's glands, and through which the molecules of odour must pass before they can reach the receptors themselves. These receptors are carried on the dendrites of the first-order olfactory neurons, the axons of which make contact with the second-order neurons in the olfactory

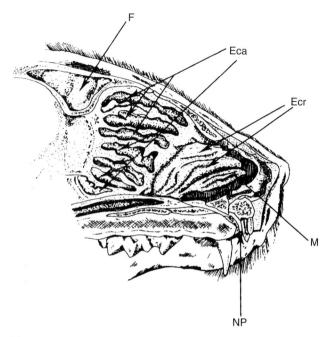

Fig. 2.10. Diagram of the bony structures that support the olfactory apparatus in the cat. The olfactory epithelium is attached to the caudal ethmoturbinals (Eca). Other structures shown are the frontal sinus (F), the cranial ethmoturbinals (Ecr) and the maxilloturbinals (M). The opening of the nasopalatine canal, which leads to the vomeronasal organ, is also shown (NP).

bulb. The detailed structure of the part of each receptor cell that projects into the mucus, the vesicle, gives some further indications that the cat's sense of smell is highly sensitive. Most of the individual receptors are carried on ciliary processes that project sideways into the mucus, up to 150 per vesicle and up to 80 microns in length, much greater than those of most other mammals. Between the receptor cells are the supporting cells, which send unusually long, abundant microvilli into the mucus, presumably as mechanical support for the cilia. There are several hundred kinds of receptor protein, but it is still not clear precisely how these encode the many thousands of odours that can be discriminated (Zarzo, 2007).

The olfactory abilities of cats have not been extensively studied, but it is likely that they are similar to those of the domestic dog, which is much more readily trained to carry out the appropriate psychophysical discriminations. Dogs are capable of detecting some compounds at thresholds between ten and 100,000 times lower than man's, and can discriminate between identical twins on the basis of odour alone. Cats almost invariably investigate novel objects by sniffing, and communicate by a variety of olfactory signals (Chapter 5, this volume); the size of the olfactory bulbs and the olfactory membrane, as well as the structure of the olfactory receptors, all argue in favour of a considerable reliance on the sense of smell.

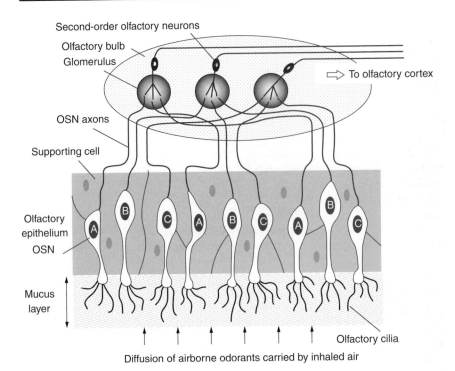

Fig. 2.11. Organization of the mammalian olfactory system. Volatile molecules reach the main olfactory epithelium in the nasal cavity, diffuse through the mucus layer and interact with olfactory receptors (ORs) in the fine cilia at the terminal knobs of the olfactory sensory neurons (OSNs). Three populations of OSNs (A, B, C) are depicted, each expressing only one of the approximately 1000 OR genes. Axons from all OSNs expressing the same receptor gene converge on one or a few glomeruli in the olfactory bulb, where they synapse with the dendrites of second-order neurons, which in turn project to the olfactory cortex. The almost 2000 glomeruli in the rat olfactory bulb are spherical knots about 50–100 μm in diameter, containing the incoming axons of OSNs, the apical dendrites of second-order neurons and the dendritic and axonal processes of interneurons (from Zarzo, 2007).

The vomeronasal organ

The vomeronasal, or Jacobson's, organ occurs in many mammalian species, but not in higher primates, including man. Their structure suggests that they are used only intermittently, as accessory olfactory organs. They are connected to both the oral and nasal cavities by the nasopalatine canal, which is about 15 mm long and runs through the incisival foramen; the lower opening can be seen as a slit immediately behind the upper incisor teeth. The paired vomeronasal organs are blind sacs running backwards from the canal, to which they are connected by very fine ducts, only 30–40 μm wide. Thus the penetration of odours to the chemoreceptors in the organs themselves is unlikely to occur passively, in the way that odour molecules can reach the olfactory epithelium every time that the cat breathes. It has been suggested that a vasomotor pumping mechanism expels some of the fluid that fills the sacs out into the canal, and then sucks it back again, drawing in odorants from the mouth and nasal cavity (Salazar *et al.*, 1996).

The mammalian vomeronasal organ is equipped with three 'families' of receptor proteins – the V1Rs, the V2Rs and the FPRs, of which only the first has been studied in cats. The V1Rs vary considerably between taxa (Young *et al.*, 2010); the platypus and mouse have over 200 variants, while the cat has 30 and the dog only 9 (humans have 3), the number presumably reflecting the discriminatory power of the organ and therefore its importance to the sensory milieu of each species. Dogs have no functioning V2Rs, so it is probable that cats also have none.

The external sign that a cat is using its vomeronasal organ is the gape or 'Flehmen' response, a 'grimace' in which the upper lip is raised and the mouth is held slightly open for a few seconds. This is performed by both males and females, in heterosexual encounters mostly by males, following actual naso-oral contact with urine scent marks or females. Females will respond in the same way to urine marks, if there is no male present (Hart and Leedy, 1987). The need for actual contact with the olfactory material implies that this is a sense more akin to taste than to smell, because the stimuli may be fluid-borne throughout. The only stage at which they might be in the vapour phase is in the transfer from the nose and lips to the opening of the nasopalatine canal. The precise role of the gape in sexual behaviour has not been fully investigated. In other species it has been found that a fully functional vomeronasal organ is essential for successful completion of the first courtship sequence, but that sexually experienced animals can rely on the olfactory sense alone to identify oestrous females.

Conclusion

Many of the differences in sensory abilities between cat and man can be ascribed to the cat's exclusively carnivorous lifestyle; communication usually evolves to suit existing sensory abilities rather than vice versa. The cat's prey may be first detected either visually or by sound. Its vision functions at much lower light intensities than our own, and is also more highly tuned to the detection of rapid movements. In order to achieve these specializations, the cat has sacrificed a certain degree of visual detail and the ability to discriminate between the thousands of colours that humans can see. The ears of the cat are very sensitive to frequencies that we regard as ultrasound, and it is unlikely to be a coincidence that these are the frequencies that are used by small rodents for communication. In almost all other respects, man has the more discriminating hearing, probably associated with our need to distinguish between the subtle inflexions of speech.

The approach to prey is aided by the cat's superb sense of balance. Although vision is probably the main sense guiding the pounce, once the cat is actually in contact with its prey its defective close focusing will rule out vision for guiding the kill. At this stage the dominant sense is likely to be that of touch, for which the whiskers, pushed forward in the pounce, the face and the forepaws are particularly sensitive. The integration of these receptors with the killing bite will be described in the next chapter.

The only sense organ that can be categorically associated with social behaviour is the vomeronasal apparatus. Flehmen, the outward sign that the vomeronasal organ is being used, is never seen in connection with feeding behaviour. Olfaction, which is much more sensitive in the cat than in man, is used when the cat is deciding whether or not to eat, but also has an important role to play in social interactions. It is still by no means certain why cats, and many other mammals, have these two olfactory systems, although in the mouse the vomeronasal organ is specialized for intraspecific communication. Part of the reason must lie in the different ways in which the odorous molecules reach the receptors – passively and through a thin barrier of mucus for nasal olfaction, actively and through a thicker barrier (which presumably slows both the onset and decay of the olfactory signal) for vomeronasal olfaction.

Mechanisms of Behaviour $\quad$ 3

Introduction

The sense organs described in the last chapter bombard the cat with data at every waking moment; somehow the important information must be filtered from the irrelevant (Dukas, 2002). This process may start in the sense organs themselves, as illustrated by pattern recognition within the visual system (see Chapter 2, this volume), but much of it happens in the spinal column and the brain. The translation of sensory input into what we see as behaviour can occur in several ways, and at different levels of complexity. If there is a direct connection between the sensory information and the behaviour pattern, as is the case in simple reflexes, the reaction time will be short, but there will be little scope for flexibility in the response. If the information is thoroughly processed by the brain before any behaviour pattern is triggered, reaction time is likely to be longer, but the stimulus–response relationship can be considerably modified by learning. Some behaviour patterns are so essential for second-to-second survival that a cat could not afford to learn them from scratch; others have to be learned because the relevant sensory information is different for every individual – for example, the learning of routes around a home range. It is important for us to try to understand the mental, as well as the sensory, capabilities of the cat, to comprehend fully the subjective aspects of the world in which the cat lives. Every cat owner has ideas of how 'clever' their animal is, but these are usually built on anthropomorphic concepts, some of which are appropriate to a carnivore, but many of which are not.

This chapter deals with the role of the central nervous system and related physiological mechanisms in determining the behaviour patterns that we observe.

© J.W.S. Bradshaw, R.A. Casey and S.L. Brown 2012. *The Behaviour of the Domestic Cat,* 2nd Edn (J.W.S. Bradshaw *et al.*)

Underlying rhythmical processes will be dealt with first, followed by the species-specific, reflexive patterns that confer some of the cat's special abilities. Finally, more complex learning will be discussed, again emphasizing those abilities and constraints which seem to separate the cat from other mammals.

Rhythms of Behaviour

Cat behaviour is influenced by underlying rhythms in the endocrine and nervous system, which are themselves affected by external patterns, such as night and day, and seasonal changes in day length. The annual rhythms have not been studied in great detail, but the hormones adrenaline and noradrenaline vary considerably with the seasons (Randall and Parsons, 1987). Even at constant temperature, food intake peaks in the autumn and is lowest in the spring, while body weight is lowest in the summer and highest in mid-winter, suggesting that metabolic rate may also be subject to annual rhythms. Cats are also affected by daily (circadian) rhythms of activity that are endogenously longer than the normal day, at about 24.2–25.0 h (Johnson *et al.*, 1983), but are reset each day by the cycle of light and dark, so that in practice they repeat every 24 h.

Sleep

The cycle of sleeping and waking is very variable, but is almost always less than 24 h long, because cats tend to sleep for several short periods during both day and night, rather than in a single sustained session. Sleep has been studied extensively in the cat, and a particular area of the brainstem, the reticular formation, is known to be a major control centre. Nerve impulses from the reticular formation to the cortex keep the cat awake; these impulses are stimulated in turn by sensory input, both direct, from the sense organs, and also via the cortex in the case of learned signals, such as the visual characteristics of a potential threat. There are other interactions – for example, hunger and thirst tend to suppress sleep, acting mainly through the hypothalamus.

The rhythmic patterns in the brain during sleep can be recorded from the skin on the head. When a cat is awake, these electroencephalogram (EEG) patterns have low amplitude but high frequency, and vary greatly depending on how active the cat is, and what it is doing. The onset of sleep is marked by a change to a high-amplitude but much lower-frequency EEG, with occasional bursts of medium-amplitude, intermediate-frequency waves. The cat then looks as if it is asleep, but is readily woken. After about 10–30 min, the EEG changes again to low-amplitude, high-frequency patterns rather similar to those of wakefulness, but the cat is now difficult to rouse; this apparent anomaly has given rise to the term *paradoxical sleep* for this phase. After another 10 min or so, normal sleep is resumed, and the two types may alternate if the sleeping bout persists. During paradoxical sleep, there is an almost complete loss of muscle tone, although individual muscles may contract suddenly, bilateral eye movements can be observed (hence this is sometimes also referred to as rapid eye movement or REM sleep) and the tail and whiskers may twitch (Oswald, 1962). All of this implies that cats in this state are dreaming, although we can have no direct evidence for this. Certainly paradoxical sleep seems to be more important than normal sleep, because the less sleep a cat has the greater is the proportion of the paradoxical phase.

The precise function of sleep (in mammals in general) still remains unresolved, even though the deleterious effects of sleep deprivation are self-evident. In kittens, the correlation between periods of sleep and both intense brain development and high synaptic plasticity suggests that sleep is indeed essential to memory formation (Frank, 2011).

Reflex Behaviour

Because the behaviour of mammals is so easily modified by experience, it is easy to lose sight of the fact that much of their minute-to-minute behaviour is controlled largely by reflexes. Before discussing the more 'intelligent' aspects of cat behaviour, some of the more pre-programmed patterns will be described. Many of these do not fit into the definition of a simple reflex, which is a brief, stereotyped motor output produced by a standardized input to peripheral receptors, acting by way of relatively simple nervous connections. One example of a simple reflex is the scratching response to irritation on a particular point on the skin; some other examples, from kittens, will be considered in the next chapter.

As the study of neurophysiology has become more sophisticated, it has been possible to study patterns of behaviour that are controlled by quite complex, interacting nervous connections; for the sake of simplicity, these will also be considered as reflexes. Many characteristic behaviour patterns can be considered as complex reflexes, because their form is largely independent from input from the forebrain, the part of the brain that processes much of the sensory information, and is responsible for most learned and 'conscious' behaviour. These include most aspects of locomotion, including walking and climbing, and the characteristic postures for urination and defecation, as well as the burying of excreta. The latter, for example, can occur with little or no feedback from the senses, as when a cat, after using a small litter tray, performs stereotyped scratching movements in thin air around the tray. Some patterns of agonistic behaviour, such as dilation of the pupils, piloerection, hissing, growling, tail-lashing and protrusion of the claws, are also reflexive, although others, such as arching of the back, biting and striking out with the forelimbs, are elicited from the hypothalamus in the forebrain (see below). Similarly, some components of oestrous behaviour appear to be reflexive, including rubbing, rolling and calling, the oestrous crouch and treading with the hind legs, as well as parts of the after-reaction following mating (Bard and Macht, 1958).

Posture-maintaining reflexes

Postural control in the cat is maintained by two feedback systems, in addition
to an anticipatory system that acts on input from the brain as soon as the cat
'decides' to change its position (Deliagina *et al.*, 2006). The simplest feedback
system is in the trunk, where two closed-loop systems, one in the spinal cord
and the other in the brainstem and cerebellum, correct for involuntary changes
in posture, such as when one foot slips on a narrow perch.

The second feedback system incorporates information coming from the
head, specifically the vestibular (balance) and visual systems. The information
produced by the balance organs has already been described (see Chapter 2, this
volume); it is relayed both to the cerebellum in the brain (see below) and also
directly to some sets of muscles to form reflexes. Of these reflexes, the simplest
are those that trigger contractions of the muscles that direct the eyes, because
eye movements do not bring about changes in the orientation of the head to
the body, and therefore do not in themselves trigger further signals from the
balance organs. These vestibulo-ocular reflexes allow the gaze to be fixed while
the head is moving slightly. As the head swivels, the direction and extent of the
rotation is picked up by the semicircular canals and is translated into an exactly
equal and opposite rotation of the eye. More prolonged turning, in which it
would be impossible for the eyes to remain fixed on one point, results in intermittent repositioning of the eyes through repetitive compensatory movements
known as nystagmus. These are rhythmic movements of the eyes, consisting of
a slow deviation in one direction, matching the turning of the head, followed
by a quick return to approximately the original position. This reflex allows for
intermittent clear vision, whereas if the eyes were held stationary in the head,
vision would be partially disrupted for the whole of the turn.

When a cat's attention is drawn to something to one side, its eyes will move
first to look at the object, followed quickly by a rotation of the head, which must
be accompanied by a counter-rotation of the eyes if the gaze is to remain on
the object. This compensation is driven almost entirely by the vestibulo-ocular
reflex system. When a cat is moving, similar reflexes allow the gaze to be corrected for the effects of jolts and jerks due to unevenness in the terrain.

The vestibular system measures small, rapid changes in position or angle
much more accurately than large, slow movements, and for the latter the simple
reflexes described above would result in under- or over-compensation if used
alone. The matching of visual and vestibular signals probably goes on all the
time in a continuous learning process. To take an analogy from human experience, this adaptability is shown by the process of adjustment to the wearing
of strong corrective spectacle lenses. When first worn, such lenses produce an
apparently disturbed motion of the surroundings at the periphery of vision, due
to a mismatch between the vestibular signals and the altered visual field, but
within a few days these aberrations disappear.

The reflexes in the neck muscles are essentially an error-correcting system. Any rotational displacements of the head will cause the appropriate neck

muscles to be activated, such that the disturbing force is counteracted and the head is restabilized. Since any movement of the neck muscles is likely to cause the head itself to move, triggering more signals from the semicircular canals, the detailed working of these reflexes is more complex than those involving the eyes. More complex still are those that trigger contractions of the body muscles, whose effects on the orientation of the skull are unpredictable. One of the simplest of these occurs at the beginning of a fall; within 70 ms of a cat losing its footing, signals from the otolith balance organs trigger extensions of the legs, as a preparation for landing (Watt, 1976). The semicircular canals stimulate reflexes that will tend to restore body position; for example, if the head rotates to the left, both front and hind left legs are extended, while both right legs are flexed. However, in many real instances the neck reflexes will act first, and thereby complicate the extent and direction of the body reflexes (Fig. 3.1).

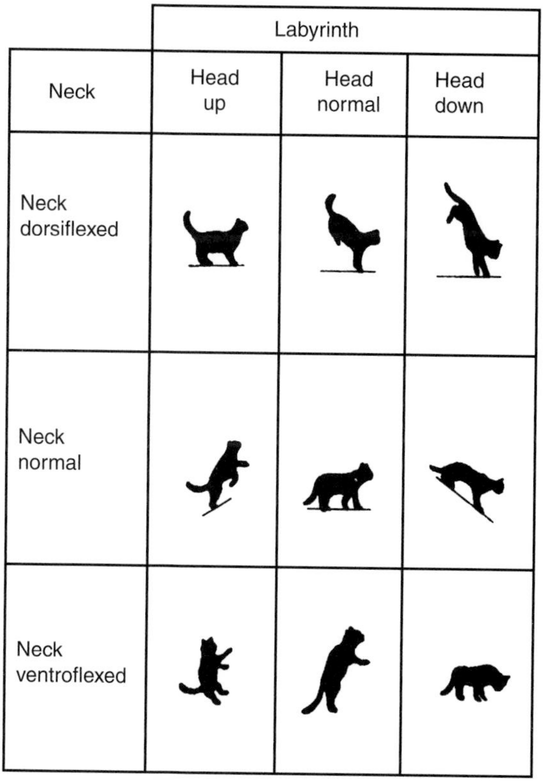

	Labyrinth		
Neck	Head up	Head normal	Head down
Neck dorsiflexed			
Neck normal			
Neck ventroflexed			

Fig. 3.1. Interactions between the static labyrinthine reflexes and the neck reflexes, and their effects on the limbs. The central figure shows the normal resting posture. In the middle row (left and right) the labyrinthine reflexes operate alone; in the centre column (above and below) the neck reflexes operate alone. Their interactions are indicated in the four corner figures. See text for further interpretation (redrawn from Wilson and Melville Jones, 1979).

Locomotion

The basic patterns of locomotion are contained in spinal 'programmes' that produce the main features of rhythmic stepping for the various gaits described in Chapter 1. The spinal cord contains specialized autonomous stepping generators for the hind limbs, and probably also for the forelimbs. Each of these contains the pattern for a complete step of a single limb, which can be speeded up or slowed down as necessary. Alternative neural pathways between these generators allow for the different ways in which the individual limbs follow one another to produce the various gaits. Signals from proprioceptors in the limbs are integrated into the stepping cycle, their effects depending upon precisely where in the stepping cycle they occur. They also allow for corrections due to, for example, uneven terrain (Rossignol *et al.*, 2006), while the whole programme is activated and sustained by command signals from the brain (Grillner *et al.*, 2008).

During locomotion, cats tend not to fix their gaze on objects in front of them, but rather look downwards towards the ground, for periods of about 0.25 s at a time before shifting their gaze, sometimes blinking at the same time (Fowler and Sherk, 2003). Thus most of the visual information that they gather while moving is itself moving across their visual field. Cats are perfectly capable of tracking fast-moving objects by moving their heads, so they can presumably therefore do the same while moving. Keeping the head steady during locomotion appears to be essential for paw placement, since if the normal visual flow is disrupted by illuminating the ground with a strobe light, cats frequently tread on obstacles that they can easily avoid under normal illumination. Motion-sensitive neurons must therefore play a major role in foot placement.

Because they don't usually look at their feet while walking, cats must memorize two to four steps' worth of the ground in front of them. If they're distracted, this visual memory persists for only 2–3 s, consistent with being stored only in working memory. However, the memory that results from actually stepping over an obstacle with the forelegs can last for up to 10 min, as shown by the corresponding movement of the hind legs when the cat is distracted while the obstacle is beneath its belly (McVea and Pearson, 2007).

The orienting reflex

Cats, like most mammals including man, will rapidly orientate their sense organs towards any sudden event in the environment. This involves complex movements that are situation-specific and therefore far from rigid, so the term reflex is used here in its sense of the behaviour occurring very quickly after the onset of the stimulus. The motor patterns involved are not specific to either the quality or intensity of the stimulus, which can be provided by any one of the senses, or a combination. The most important features of the stimulus are its contrast and novelty; the ending of a continuous stimulus such as a drawn-out sound can evoke the same response as the beginning of that sound. If the same stimulus is repeated over and over again, the reflex becomes weaker and is finally not elicited at all. In the brain, one major effect is the dilation of the cerebral blood vessels and constriction of the peripheral, which facilitates the transmission of information through the central nervous system, making the cat more 'attentive' (Sokolov, 1963). The essentially pre-programmed nature of this reflex can be illustrated by its invariant effects on the eyes. If an object appears suddenly in the visual field, the pupils dilate and the eyes automatically focus at their shortest possible distance, even if the object is actually far away. Non-visual stimuli have exactly the same effect on the eyes, whether they are odours, sounds or a light touch, always provided they occur with an element of surprise.

Grooming

Cats spend a great deal of their time grooming; of the half of their lives that is not devoted to sleep, oral grooming, supplemented by the occasional bout of scratching, can occupy as much as 10%. Cats' tongues incorporate cornified papillae that are specialized for cleaning the fur as they lick, and the small incisor teeth are also used. Although grooming is effective at removing ectoparasites such as fleas, it appears to be regulated by a programmed grooming generator, both in its timing and its typical form (Eckstein and Hart, 2000), rather than always being prompted by itching at a specific location on the skin. Cats often groom within a few minutes of waking, suggesting that this generator is 'catching up' after an enforced period of inactivity. The typical cephalocaudal sequence of grooming, starting with face-washing using the paws and then progressing to hind legs, flanks, neck and chest, anogenital area and then the tail, also suggests internal, rather than stimulus-driven, programming.

The Brain and the Control of Behaviour

The brain, and particularly the forebrain, exerts a controlling influence at almost every stage of the more complex reflexes. For example, the reticular formation in the brainstem not only controls sleep, but also the general state of arousal. It influences the impact of all the sensory systems on the cerebral cortex, and is particularly active during habituation, the process whereby the same stimulus, if repeated, elicits a weaker and weaker behavioural response. A second, parallel arousal system in the mid-brain mediates the effects of learned behaviour patterns (Colgan, 1989).

In addition to such non-specific effects, it has proved possible to group some behaviour patterns together, based upon the site in the brain from which they originate. One of the best understood is the 'quiet biting attack', which is the psychologists' term for the patterns seen in the latter stages of hunting, culminating in the kill. Groups of neurons in the hypothalamus and mid-brain control a whole sequence of events, each one of which contains several reflexes. In the order in which they occur, these are:

1. Stalking, sniffing and visually guided approach to the prey.
2. Visually guided orientation of the cat's head to the target, assisted by tactile stimulation from the forepaw if this makes contact with the prey.
3. When the head reaches the target, precise orientation of the snout by tactile stimulation of a trigger zone on the face.
4. Opening of the jaws, in response to stimulation of a trigger zone around the lips.
5. Closure of the jaws when a trigger zone just inside the mouth is activated.

The hypothalamus has an important role to play in changing the thresholds for the component reflexes. For example, the seizing and biting reflexes are switched on, while others that would interfere with the capture of prey, such as the jaw drop reflex, are suppressed. The sensory inputs required at each stage can be defined precisely; the head-orienting reaction occurs in response to touch over an area of skin extending from just above the upper lip to the hairless area on the nose, and out to the side as far as the whiskers. Biting requires a touch on either the upper or lower lips, most effectively at the front of the mouth, but to a lesser extent around to the sides (Fig. 3.2). Persistent biting requires stimulation of the trigeminal receptors around the mouth, as well as touch receptors (Siegel and Pott, 1988).

Quite separate areas of the hypothalamus and mid-brain (specifically, the dorsal half of the periaqueductal grey) and the amygdala control a group of defensive behaviour patterns, including retraction of the ears, piloerection, arching the back, marked dilation of the pupils, vocalizations and unsheathing of the claws (Siegel and Shaikh, 1997; McEllistrem, 2004). Yet other areas of the brain control flight behaviour. Thus the way that many species-specific patterns are organized in the brain mirrors the groups in which we can place them, based on their functions in free-ranging animals.

Comparisons with other species

We can also deduce something of the special features of the cat's brain by making comparisons with other species. One concept that has proved successful in

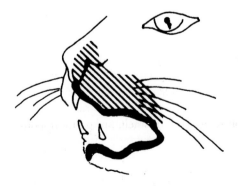

Fig. 3.2. Sensory fields that trigger the head-orienting (shaded area) and jaw-opening (solid areas) responses during prey-capture. The lower jaw also contains a less well-defined area (not shown) directing the head-orienting response (redrawn from Macdonnell and Flynn, 1966).

such comparisons is that of structural encephalization, which is defined as the enlargement of the brain beyond that expected from the size of the body, and is measured as an encephalization quotient (EQ) (Jerison, 1985). Large bodies need large brains because they have larger muscles and more extensive somatic sensory systems, but once this is allowed for some striking comparisons can be made. For example, deer, wolves, crows and lemurs all have roughly the same EQ, while hedgehogs have retained the lower EQ of the earliest mammals. This measure contains an element that could be defined as 'intelligence', but enlargement of the brain as a whole can also be due to specializations, such as an increase in the sensitivity of one or more of the senses, which will produce more information for processing. Some of the ecological factors that have been proposed as requiring an increase in EQ are movement in three dimensions (flying, swimming) compared with two (terrestrial); an active anti-predator strategy compared with a passive one (e.g. the hedgehog); and a long period of parent–young association for the transfer of skills (Shettleworth, 1984). More nebulous, but intuitively correct, is the idea that learning abilities differ between animals in terms of how flexible that learning can be. Higher primates can learn a wide variety of tasks and associations; the learning abilities of lower mammals seem to be more situation-specific, more constrained by the niche for which they have evolved. The former is likely to raise the EQ more than the latter. Thus EQ is built up from several components, which can have very different emphases in different species; overall the figure reflects an investment in information processing power, whether it be for learning, or for a special skill such as social cognition, or for sensory ability. The first of these is the most flexible, the latter two are likely to be more niche-specific.

The brain of the domestic cat is very similar to that of other members of the genus *Felis*. The basic pattern appears in the fossil record some 5–9 million years ago; the brains of earlier cats, most of them sabre-toothed (the paleofelids), appear to have been organized along different lines (Radinsky, 1975). Two of the most striking features of the brain of the domestic cat are the enlargement of the cerebellum, coordinating balance and movement, and the large proportion of the cortex devoted to controlling movement; cats' brains reflect their athletic prowess. The part of the cortex that deals with hearing is well developed, but the olfactory bulbs are, compared with those of other carnivores, rather small. The felids as a family have rather little space for large olfactory bulbs in their comparatively short skulls; those that do have large olfactory bulbs, like the lion, have large home ranges, a trend repeated across all the Carnivora (Gittleman, 1991). The visual area of the cortex is less developed in the domestic cat than in some other felids, for example the jaguarundi, *Herpailurus yaguarondi*.

The cat's EQ is higher than for the majority of rodents, and about average for the carnivores. The dog family has the highest average EQ of any carnivore, some 25% higher than for the average felid; larger olfactory bulbs in dogs, reflecting a greater reliance on their sense of smell, are partly responsible, but the prefrontal cortex, which is thought to selectively inhibit primitive behaviour patterns, is also larger in some dogs. It has been suggested that such inhibition, for example the substitution of aggressive by submissive behaviour patterns, may be a component of the complex social behaviour seen in wolves and other canids.

Learning and Intelligence

It is impossible to draw a sharp distinction between instinctive and learned behaviour in an animal as complex as the cat. Species-specific behaviour, such as vocalization, mating behaviour, some aspects of hunting and the reflexes displayed by kittens, are presumably based upon inherited patterns, but these are modified, supplemented and altered, in both the long and the short term, by learned components. Although the cat has been a favourite subject for the study of learning itself for more than a century, many of the earlier studies were conducted on indoor-reared animals, which are less adept at learning visual discrimination tasks than are cats that have grown up outdoors (Žernicki, 1993). Moreover, many have taken insufficient account of the evolutionary pressures that have shaped the mind of the cat, compared with those that have shaped, for example, that of the rat or chimpanzee. While the vast majority of species are closely adapted to their current niches, their abilities to respond to sudden changes in those niches vary considerably. Animals relying largely upon instinct, or highly context-specific learning, will only be able to re-adapt at a pace determined by evolutionary mechanisms. Those with more extensive learning abilities can alter their behaviour patterns rapidly; they possess a capacity to solve problems by systems that have not been selected specifically by adaptation to current niches, but are available to cope with unexpected change.

The domestic cat seems to be a prime example of an adaptable species, given that it is able to move from total dependence on man to semi-independence and back, within a lifetime or at most a few generations. Such abilities are, almost by definition, not straightforward to assess, since their full value will only be expressed under circumstances of rapid change in the environment. However, certain mental skills, such as learning by imitation, and the formation of mental concepts, are likely to contribute to the flexibility required, and these are described in the section below on complex learning. On the other hand, cats are not infinitely flexible, for there is ample evidence that their learning abilities are species-specific at all levels of complexity, including straightforward associative learning.

Learning by association

At its simplest level, learning involves the linking together of previously unrelated stimuli, or between actions and the consequences of those actions. Even invertebrate animals are capable of this type of learning, and so it is hardly surprising that cats can form a wide range of associations of this type. Indeed, in the past the behaviourist school of animal psychologists has attempted to describe most learning at this level, but it is evident that many mammals, including cats, are capable of much more complex mental processes, some of which will be described in the next section. By contrast, popular accounts of cat behaviour tend to express learning in the terms of human subjective experience, almost as if cats were mentally defective people rather than highly adapted carnivores. Such controversies are far from new, as the following excerpt from Hobhouse (1915) will illustrate:

> I once had a cat which learned to 'knock at the door' by lifting the mat outside and letting it fall. The common account of this proceeding would be that the cat did it in order to get in. It assumes the cat's action to be determined by its end. Is the common account wrong? Let us test it by trying explanations founded on the more primitive operations of experience. First, then, can we explain the cat's action by the association of ideas? The obvious difficulty here is to find the idea or perception which sets the process going. The sight of a door or a mat was not, so far as I am aware, associated in the cat's experience with the action which it performed until it had performed it. If there were association, it must be said to work retrogressively. The cat associates the idea of getting in with that of someone coming to the door, and this again with the making of a sound to attract attention, and so forth ... Such a series of associations so well adjusted means in reality a set of related elements grasped by the animal and used to determine its action. Ideas of 'persons,' 'opening doors,' 'attracting attention,' and so forth, would have no effect unless attached to the existing circumstances. If the cat has such abstract ideas at all, she must have something more - namely, the power of applying them to present perception. The 'ideas' of calling attention and dropping the mat must somehow be brought together. Further, if the process is one of association, it is a strange coincidence that the right associates are chosen. If the cat began on a string of associations starting from the people in the room, she might as easily go on to dwell on the pleasures of getting in, of how she would coax a morsel of fish from one or a saucerful of cream from another, and so spend her time in idle reverie. But she avoids these associations, and selects those suited to her purpose. In short, we find signs on the one hand of the application of ideas, on the other of selection. Both of these features indicate a higher stage than that of sheer association.

Hobhouse evidently interprets his cat's behaviour as having purposeful elements. However, he does go on to offer an alternative explanation, which we would now class as behaviourist. This is based upon an association between the 'pleasure' of getting through the door and the action of lifting and dropping the mat, and no more. The action assimilates the character of its result and becomes in itself attractive to the cat.

Pavlovian learning

Even with such a simple sequence of events as this, psychologists do not fully agree on the precise details of the learning mechanisms involved. Their findings, usually based on simple tasks carried out under highly defined conditions, are not always easily interpreted in functional terms (i.e. the value to the cat of the type of learning observed). One of the simplest forms of learning is known as Pavlovian conditioning, named after the classic experiments of Pavlov, who trained dogs to anticipate the arrival of food in response to arbitrary signals, such as the sound of a bell or metronome. The dog has continued to be a favourite subject for this type of study, so some of the examples to be described, although drawn from the dog, will be assumed to apply to the cat also. The primary function of Pavlovian learning seems to be the acquisition of information about stimulus relationships in the environment. One stimulus, the unconditioned stimulus (UCS), is normally linked to a particular motivational state, and releases an innate reaction, the unconditioned response (UCR); for example, the smell of food will result in salivation in a hungry animal. If a second stimulus, the conditioned stimulus (CS), occurs consistently with or immediately before the UCS, it will come to release the UCR even on its own; the UCR has become a conditioned response (CR). However, the UCR and CR need not be identical, although they are normally linked to the same type of motivation. For example, if the UCS is the pain inflicted in an attack by an aggressive tomcat, the UCR will most likely be flight. The victim will probably associate the appearance of the attacker (the CS) with the experience of the fight, and will respond appropriately (CR) on seeing the attacker again. However, this response may change with circumstances; if the CS is fairly distant, the CR may be to 'freeze' in an attempt to avoid detection, rather than to run away, inviting a chase (an example of an interaction between Pavlovian and instrumental learning). Such considerations, and more direct experiments, have led to the conclusion that in most cases the association built up in Pavlovian conditioning is genuinely formed between the CS and the UCS, and not between the CS and the CR; in ethological terms, if the UCS is a releaser, the CS is a learned releaser, bringing it under the control of normal motivational systems.

The interspecific differences that this can produce have been illustrated by a comparison between the learned feeding responses of cats and rats (Grastyan and Vereczkei, 1974). The arrival of a food reward was signalled by 10 s of a clicking sound coming from a loudspeaker 2 m away from the food store. This combination caused the cats to run towards the sound, and some would search all around the loudspeaker, and even attempt to bite it. When this response was most intense, the cat would often not take the food reward at all, although after hundreds of further trials the feeding response was re-established. Under similar conditions rats would briefly turn their heads towards the sound, but would rarely approach it. For the rats, the sound was an initially irrelevant cue, but for the cats, which use auditory cues extensively while hunting, it was not, and evidently some conflict appeared between the apparent location of the 'prey', as indicated by the sound, and its subsequent appearance as food.

An important feature of the relationship between CS and UCS is that they must be contingent; if the pairing is unreliable, the response (CR) is considerably weaker than when the CS and UCS always occur together. This prevents the cat from acquiring false or poorly predictive information about its environment. Events that reliably do not predict the UCS are also learned, as can be shown in two separate ways. First, if a CS is repeatedly presented in a way that does not predict the arrival of the UCS, then when the same CS is presented with the UCS it is difficult to establish the connection; the cat has already learned that the CS is an irrelevant cue, and so when its prediction value changes there is a delay before the new association is registered. Secondly, if the procedure described is performed in reverse, the association is rapidly 'unlearned' once it is no longer predictive. Furthermore, associations can be learned between two neutral stimuli (i.e. not releasers) that reliably occur together, even when neither brings about any overt behaviour. This can be shown by pairing just one of the two stimuli with a UCS, after which both stimuli will release the CR. This 'behaviourally silent learning' is of obvious value to, for example, a cat learning the topography of its home range, although cats also possess more advanced orientational abilities, which will be described in a later section.

Pavlovian learning is probably the basic mechanism behind many other behavioural phenomena, including taste aversion learning (discussed in Chapter 6, this volume) and some aspects of foraging behaviour. For example, if cats form the equivalent of the 'search images' used by birds to detect cryptic prey (Zentall, 2005), they may do so by associations between the appearance of a specific prey type and its profitability. Specific features of the environment may come to be associated with particular prey types or prey densities. Moreover, Pavlovian associations can also prepare the cat for subsequent events so that they can be optimally exploited; for example, CSs indicating food bring about physiological changes that speed up digestion once the food is actually eaten.

Instrumental learning

These simple Pavlovian mechanisms should enable a cat to build up a much more organized picture of its world than would instinct alone, but they will not on their own produce the flexibility in behaviour that cats are evidently capable of. For the latter, a different type of learning is required, one that will enable the cat to predict the consequences of its own actions, and modify those actions based on past successes and failures. This is addressed by the psychological technique of instrumental learning, in which the subject has to respond in some way to a stimulus; correct responses are rewarded. Some of the earliest work in this area used cats as subjects, particularly the puzzle-box experiments of Thorndike (discussed in Hobhouse, 1915). Thorndike placed cats in cages from which they could escape by means well within their motor capabilities, such as clawing at a string, depressing a lever, pushing aside a swing door, and so on (Fig. 3.3). The cats would claw and scratch indiscriminately at the sides of the cage, until by accident they performed the right action and gained their freedom. The time that it took the cats to escape declined with repetition, implying that the probability of performing the correct action was increased by each success. Some of the tasks set were quite complex; one latch required a simultaneous lift and push, and in other cages two or even three latches had to be opened in the correct sequence. Although not all cats could master these, all were opened by some. Taking an average of several

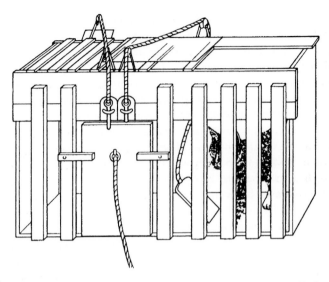

Fig. 3.3. An example of one of Thorndike's puzzle boxes (from McFarland, 1985).

animals, the skills appeared to be gained gradually, and Thorndike concluded that 'The gradual slope of the time-curve, then, shows the absence of reasoning. They represent the wearing smooth of a path in the brain, not the decisions of a rational consciousness.' However, individual animals did not behave quite in this probabilistic way. Some did take a little less time to escape on each successive attempt, but many seemed to improve their performance quite abruptly, and then never make another mistake, even with an interval of several months between trials. In fact, rapid (one-trial) learning is nowadays not thought to be good evidence for conscious thought. Many animals learn crucial associations, such as the toxicity of particular foods, after only one experience; in this situation the potentially lethal consequences of eating the same food again are likely to outweigh the risk that, after only one encounter, the animal has learned the wrong connection.

Thorndike's idea that random behaviour patterns were shaped by successes gave this type of process its alternative name, 'trial-and-error learning'. The apparently random behaviour of the cats when first put into the cages, together with the results of many other such experiments on other species, helped to establish the notion that almost any behaviour could be shaped in this way, minimizing the value of instinctive behaviour. However, it is now clear that species-specific behaviour patterns have a clear role to play in providing the behaviour that is to be shaped, in directing the attention of the cat towards the task to be performed, and in providing an assessment of the value of the reward for correct performance. These *species-specific constraints* presumably ensure that in the real world, outside the narrow context of the typical instrumental learning paradigm, the most ecologically functional skills are acquired. Thus it is much easier to train a cat to obtain a food reward by using a normal component of hunting behaviour, such as hooking back a bolt with its paw (the movement used to dislodge prey that takes refuge in a crevice), than by some arbitrary but straightforward action, such as pushing an identical bolt inwards. In the cage experiments, Thorndike found that certain actions of the cat could not be trained; for example, if the cat dislodged the latch by accident with its tail, it did not appear to learn anything about the location of the latch or the type of actions likely to open it on subsequent trials. Also, if the cat was allowed to escape every time it performed some arbitrary action, such as grooming, the frequency of grooming did not increase; the connection between the action (grooming) and the reward (escape) was never made. The type of incentive is also important. For young cats, although food is a powerful reinforcer, other activities, such as manipulation of simple objects like a ball or a crumpled piece of paper, or exploration of an unfamiliar space, are also adequate rewards for a discrimination task (Miles, 1958).

Instrumental learning methods have been used extensively to probe the sensory and mental capabilities of cats; examples of the former have been described in Chapter 2, and the latter will be explored in more detail in the following section. To provide information on the way that the cat makes every-day decisions about its actions, more complex schedules of reward and response are required. For example, one common procedure is to reward two distinct responses simultaneously, either pairing each with a reward of different value or rewarding different proportions of the two responses. In many species, the strategy that is adopted can depend on the type of stimulus (e.g. visual or spatial) used to indicate the reward, presumably reflecting species-specific propensities to learn links between food and its sensory characteristics. The types of strategy adopted by the animal can give some idea of the way it might behave in the field when confronted with prey items of different nutritive value, or that are more or less easy to catch. One set of trials that mimicked the latter situation (Warren and Beck, 1966) can be used to illustrate the procedure and its possible results. Cats were rewarded intermittently for choosing one of a pair of visually distinct wooden blocks (for example, one triangular and black, and the other circular and white). If only one block was rewarded (reinforcement ratio 100:0), the cats rapidly learned to respond only to that block. If both were rewarded equally (50:50), responses were, on average, also equal. A 90:10 ratio resulted in all cats responding in the same way as to the 100:0, a strategy known as *maximiz-ing*, because by this means the maximum number of food rewards is obtained within a session. The more complex reinforcement ratios of 60:40, 70:30 and 80:20 produced some very individual-specific shifts in strategy. Particularly at the lower ratios, most cats distributed their responses within 5% of the reinforce-ment ratio, a strategy known as *matching*. The ratio at which each cat switched strategies varied considerably, some maximizing at 60:40, others matching up to 80:20. The mechanism behind matching seemed to be a simple one; most cats persevered in responding to the stimulus that had produced food on the previous attempt, only switching to the other when this prediction failed to pay off. The origins of the individual differences could not be determined, but may have been due to the cats' previous learning experiences.

It is difficult to extrapolate from the results of such trials to real foraging decisions, because they still contain an element of artificiality, in that the cats obtain a great deal of their daily food intake away from the training procedure, so their motivation may not be the same as if they were foraging. In other species, closed-economy experiments, in which animals have to do work for all their food, have produced some conclusions that are opposed to those obtained from trials like the one described above (Davey, 1989). Very few trials of this type have used cats, but there is some evidence that the maximizing strategy can be subservient to a direction-stable strategy in which each cat has a preferred foraging route (see Chapter 7, this volume).

Various extensions of instrumental learning are used when cats are taught to do tricks (McGreevy and Boakes, 2007). It is sometimes claimed that cats cannot be taught tricks, but what is usually meant by this is that cats cannot easily be taught by the same methods as dogs. Most dogs are very attentive to their trainers, and can be rewarded by positive social contact alone. Cats are much less likely to be interested in the training process for its own sake, and usually need to be rewarded with food or other incentive such as play. The sooner the reward is given after each correct response, the more easily cats will learn the correct association. Instrumental learning tests also show that cats, in common with most mammals, remember tasks for much longer if they are only rewarded for a proportion of correct solutions (an intermittent reinforcement schedule): introducing intermittent rewarding once a behaviour is learnt can be used to fix the results of such training. These simple techniques can be used to reinforce normal behaviour patterns, which can also be trained to be shown on cue by only reinforcing responses shown after the cue is presented. Patterns that are not entirely natural can be trained by progressively rewarding responses closer to the desired behaviour, known as *shaping*. To take a simple example, cats will not usually jump over an obstacle if they can walk round it. To train a cat to jump on request, it can first be rewarded for walking over a stick that is lying on the ground, then for stepping over it when it is raised slightly. As the stick is raised further, the cat is rewarded for jumps only. Once the habit has been established, it can be made more persistent by rewarding only a proportion of successes. More complex tricks often have to be built up a step at a time using gradual shaping of responses.

Because the timing of reinforcement is important in establishing a desired response, a *conditioned reinforcer* is sometimes used to avoid delays associated in providing a reward. Conditioned reinforcement is where a previously neutral signal (e.g. a sound) is reliably associated with a primary reinforcer, for example food. Once this association is established, the sound can be used to signal a correct response, even where the trainer is at a distance from the cat, and any delays associated with getting out and providing a food reward are avoided. With patience, cats can be trained to show complex behaviours involving quite long chains of shaped behaviours. However, training sessions generally need to be short and the reward valuable to maintain interest in the task.

Complex learning

The cat is no longer a favourite subject for the study of learning – much more is known about the specific abilities of pigeons, rats and monkeys – and so the account that follows is by no means a complete description of feline intelligence.

Complex stimuli

Ecologically meaningful cues are rarely simple; they may differ from their background, and other less relevant cues, in several ways, for example size, shape, brightness, colour, characteristic movements, sounds and odours. A great deal is known about the ability of cats to identify stimuli that differ in only one sensory dimension, but much less about the analytical processes they use when confronted with complex stimuli. Some idea of these processes can be gained from experiments carried out to detect the more relevant of a pair of cues presented simultaneously (Mumma and Warren, 1968). Three-month-old cats were trained to distinguish between rectangles that differed in both orientation and brightness, and were then tested to see which one they preferred of pairs of rectangles that differed in either orientation or brightness (Fig. 3.4). These preferences showed that both cues had been learned simultaneously by most kittens, although there was the expected variability in accuracy; there was no evidence that some had relied more on one than the other, as rats tend to do. The relevance of

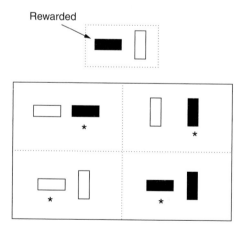

Fig. 3.4. One example from the sets of discriminations that show that cats can learn two attributes of a visual stimulus simultaneously. Young cats were rewarded for responses to the left-hand figure in the upper box, and once trained were tested for preferences between the four pairs in the lower box. Each of these differed in either brightness, or orientation, but not both. The cats tended to make choices (starred) indicating that they had learned that the rewarded object had been both shaded and horizontal (redrawn from Mumma and Warren, 1968).

particular shapes has also been examined, and one cue that cats seem to pay great attention to is whether figures are open or closed. For example, they learn to discriminate circles from U-shapes much more quickly than from triangles (Fig. 3.5). The basis for this seems to be the ratio between the area of a shape and the number of sides that it has (Warren, 1972). When irrelevant cues are presented along with relevant ones, cats are better than rodents at singling out the predictive one. For example, from a set of wooden triangles and circles that could be black or white, and one of two sizes, only triangular or circular shapes were rewarded with food. The cats learned that it was the shape that was the discriminating feature, and there was no difference in the speed of learning between cats trained with pairs of triangles and circles which were always the same shade and size, and cats trained with pairs of triangles and circles the shade and shape of which changed from session to session (Warren, 1976). The errors that cats make in such trials seem to originate in their initial preferences and aversions, which they rarely overcome as completely as monkeys do when trained on the same problems.

The concept of oddity

The precise extent to which cats can generalize from one discrimination to another is still unclear. One such generalization is that of oddity. Chimpanzees can quickly grasp the idea that they are to pick out the non-matching object in a group of three in which the other two objects are identical. Cats take much longer to learn this, and are prone to mistakes. In one set of trials, five 9-month-old cats were initially trained to discriminate the odd one (for example, a triangle) of three objects (the other two being, for example, circles), when the same set of three objects was presented at each session (Warren, 1960). Once each cat had successfully learned that it should

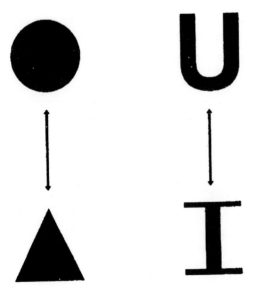

Fig. 3.5. Examples of shapes that cats find easy to discriminate from one another, the exceptions being the pairs connected by arrows.

look for the triangle, one of the circles was removed from the set and was replaced by a triangle; in other words, the oddness was transferred from the triangle to the circle. The odd object was still the one that was rewarded; initially the cats preferred either of the two triangles, because that shape had previously been associated with food, but quickly turned their attention to the circle, which was now the odd object. This reversal was repeated 20 times, and each time the cats followed the change, although one of the five was consistently more accurate than the others. This part of the procedure demonstrates that the cats could discriminate the objects from one another even when two were identical, but it does not demonstrate learning of oddity itself. In the second part of the procedure, the triangle/two circles and circle/ two triangles combinations were presented in a random order, and one cat, the best performer in the preliminary trial, learned that it was the odd one out, rather than either of the shapes themselves, that signalled food. This cat could also rapidly generalize from this pair of shapes to others; presented with random orders of two new shapes in groups of three, it made fewer and fewer errors each time the pair of shapes was changed for a new pair, showing that it had grasped that the salient cue was oddity. That the cat that mastered the oddity problems was also the best at recognizing objects suggests that these two types of learning are linked. It is also possible that all the cats had the concept of oddity, but could not be persuaded to demonstrate it by object discrimination.

Object permanence

Similar problems may lie behind apparently conflicting findings relating to the extent to which cats understand where objects have been hidden. Such skills would be highly adaptive for a carnivore hunting in cover, and so we should expect cats to be highly aware of the most likely location of prey that has gone to ground, or has moved after disappearing behind cover. A theoretical framework for the concept of object permanence, devised by Piaget for recording the development of human infants, has been used to quantify the abilities of cats and is therefore worth summarizing in its basic form. In the first two stages of development, infants show little interest in objects, and when an object is hidden, they stare at the point from which it disappeared, rather than looking round for it. Stage 3 is marked by the ability to discriminate partly hidden objects, and to recognize the part that is visible as belonging to the original whole. Stage 4 introduces the concept of permanence for the first time; objects that disappear are searched for, indicating that the infant realizes that they still exist. However, in a series of tests with the same object they tend to search the place where the object has been hidden most often, rather than the place where they have just seen it hidden; a previously successful action is repeated, akin to the result expected from instrumental conditioning. Reliance on immediate perception is established in stage 5, and more complex problems can also be solved. These include *sequential visible displacement*, in which an object is hidden in several places in turn, the solution being to look in the place closest to where it was last seen, and *single invisible displacement*, in which an object is hidden first in the hand, and then the hand is placed under a cover. When the hand is withdrawn and shown to be empty, the child should look under the cover. The final (sixth) stage completes the mental concept of object permanence, in which the child can follow *sequential and successive invisible displacements*. In the first, an object is hidden in the hand, put under one cover, shown to be under that cover, palmed again and hidden under a second cover. Infants at stage 5 tend to look under the first cover. In the successive displacement, the object hidden in the hand is moved from one cover to another, and left under the last before the empty hand is displayed; again, infants that have not reached this stage tend to start by looking under the first cover.

The extent to which cats can be persuaded to demonstrate their abilities in this area seems to depend a great deal on the protocol that is used. Cats appear to be easily fooled by the classic sequential invisible task: if a piece of food is made to disappear behind one barrier, then moved concealed inside a container first into the open and then behind another barrier, where it is made to reappear and then disappear, most cats will look behind one or other of the screens, but show no strong preference for either (Goulet *et al.*, 1994). Their behaviour suggests that they don't realize that the two objects they have seen are actually one and the same.

However, if invisible displacements are made more ecologically relevant as far as their hunting behaviour is concerned, cats do appear to be able to make predictions about where a potential prey item might have hidden itself. Using the apparatus shown in Fig. 3.6, Dumas (1992) was able to show that if a cat has to lose sight of a prey object in order to reach it, the cat does have an idea where the prey has gone if it subsequently finds that it has hidden itself. Moreover, given a choice of routes to hidden prey, cats do not always take the most direct route, possibly a tactic for confusing the prey item (Dumas, 2000). The locations where prey has disappeared are generally stored in working memory only (10–15 s) presumably because it is not worthwhile for the cat to continue to look for highly mobile prey for much longer than this (Fiset and Doré, 2006).

Concepts for orientation

Familiarity with the environment implies that cats have some kind of concept of the way that the components of the world around them fit together. This has been investigated by examining the ways that cats find their way around. There are several possibilities, incorporating different levels of sophistication. The simplest type of orientation relies on direct perception of the goal ('the rabbit warren is in the bank that I can see at the other side of the field'), or a step-by-step route based on landmarks ('if I go to the oak tree that I can see, and turn left, I will then be able to see the warren'). Many animals, including some invertebrates, use such orientation systems, which are generally simple to use but prone to error ('the oak tree has been felled, so I can no longer find the warren'). Cats rely on egocentric cues ('to my left') in simple situations where they are unlikely to lead to error (Fiset and Doré, 1996), but they are also capable of constructing cognitive maps of their surroundings, particularly if they have been able to explore them thoroughly (Poucet, 1985). Although they can construct mental maps based

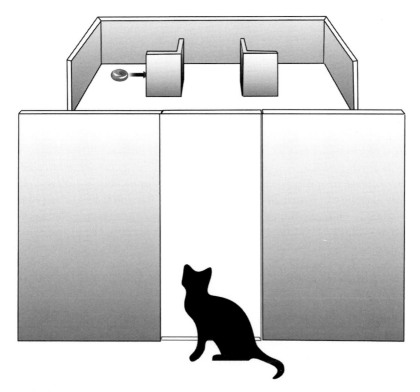

Fig. 3.6. An apparatus used to test cats' ability to follow invisible displacements. The cat starts as shown, from where it is able to see a piece of food through a transparent centre section of a screen that is otherwise opaque. Once the cat has started approaching the food, and is behind one or other of the opaque sections, the food is moved (using transparent strings) behind the inner 'hiding' screen (redrawn from Dumas, 1992).

on a brief view of relevant features, these are not remembered for more than a few minutes. Mapping leads to the possibility of taking short-cuts ('last time I went to the warren I went to the oak tree and turned left, so this time I will go diagonally across the field and through the hedge; the warren is just beyond the hedge'). It also permits the rapid choice of optimum routes; given a choice of ways to an invisible goal, cats almost always prefer the shortest one. If there are several routes of roughly the same length, the one that starts off in the direction closest to the direction of the goal itself may be preferred, a common human habit also. Minimizing the number of twists and turns in the route after that is also a factor that determines a cat's choice, but a relatively unimportant one.

Concepts of time and physics

Cats are capable of discriminating short time-intervals; they can tell the difference between a sound that lasts 4 s from one that lasts 5 s, and can also learn to delay their response to a stimulus by several seconds, again to an accuracy of about 1 s. This implies the existence of an internal clock that times the duration of both internal and external events; this could be used, for example, in assessing the rate at which particular feeding strategies produce food. Another skill which would be useful in this context is the ability to count, and it is thought that cats do have some kind of abstract conception of number, although attempts to demonstrate this have not proved recognition of numbers greater than about seven.

Cats' comprehension of causality, while little studied, seems rudimentary at best. In a string-pulling task (Fig 3.7), most cats can easily be trained to retrieve a piece of food by pulling on a single string, but seem unable to work out which of two parallel or crossed strings is attached to food and which is not (Whitt *et al.*, 2009) – dogs also fail at the crossed-strings task but can solve the parallel-strings problem.

Finally, it is worth returning to the question of how cats obtain the information on which learning is based. Trial-and-error is a time-consuming process, and in a social animal a great deal of time could be saved by watching the ways that conspecifics solve problems. Cats are certainly capable of this, even when they are adult (John *et al.*, 1968), although it has been argued that the actions of the conspecific merely help to focus the cat's attention on the problem to be solved. Some are apparently able to 'work out' exactly how to perform a task simply by watching an experienced individual carry out that task, and then repeating the actions they have seen. Learning of this kind is essential in the most intense period of the cat's social life, its life as a kitten with its mother and siblings.

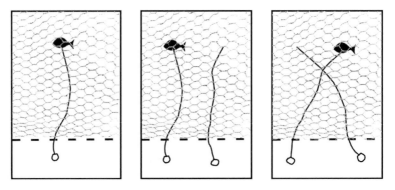

Fig 3.7. Arrangements of strings, handles and food rewards in a string-pulling task – the cat has access only to the area in front of the dashed line, the remainder being enclosed in a box with transparent sides and a mesh cover (redrawn from Whitt *et al.*, 2009). Cats can be trained to obtain a food reward by pulling the string out from the covered box (left), but when presented with two strings, one baited and the other not, seem incapable of selecting the correct string, whether the strings are parallel (centre) or crossed (right).

Behavioural Development　　　**4**

Introduction

The behaviour of a mammal is always developing, because learning is a process that starts in the womb and persists into old age. However, many of the changes we see in both young and old cats are due to a combination of learning and structural changes that occur in the sense organs and the central nervous system; maturation at the beginning of life, and degeneration at the end.

Maternal Behaviour

For the first few months of their lives, kittens spend a great deal of time with their mothers. If the mother is a pet, then there will also be a considerable influence from the human caregiver (see Chapter 9, this volume). If the mother is part of a social group, there will be considerable input from other females in the group, particularly those that produce their own litters at about the same time; communal rearing of kittens is discussed in Chapter 8. The account that follows is based on the behaviour of solitary females, and of necessity much of the most precise information has been obtained from cats confined indoors, rather than solitary outdoor cats that tend to keep their young litters in inaccessible places. Paternal influences, apart from those conveyed by genetics, are likely to be insignificant, since the pair-bond in cats is usually very weak, and in any case most queens are promiscuous given the opportunity.

A pregnant female will spend a great deal of time and effort seeking out a suitable place to give birth, and may visit and revisit several alternatives before

© J.W.S. Bradshaw, R.A. Casey and S.L. Brown 2012. *The Behaviour of the Domestic Cat*, 2nd Edn (J.W.S. Bradshaw *et al.*)

coming to a decision. The site is generally well protected, but there is little or no actual construction of a nest. The selective pressures that have shaped these behaviour patterns are not all readily apparent, because they probably evolved under different circumstances to those in which most feral cats find themselves today. Protection from the weather must be an important factor in temperate climates, where kitten mortality due to respiratory and enteritic viruses is high. Potential predators of young kittens include feral dogs, possibly some birds and other cats, although the impact of the latter is still controversial (see Chapter 8, this volume). After parturition the mother may become aggressive towards cats and dogs that she had previously tolerated. The kittens may be transferred to a new nest site even while they are still helpless; the reasons behind such moves are not fully known. The fouling of the original nest with the remains of prey and excreta does not seem to be an important factor as once thought, but transfer may serve instead to keep litters hidden from predators or to increase proximity to food sources as the kittens develop (Feldman, 1993). Some of these moves bring about the pooling of two or more litters, or occasionally the division of a joint litter, and so may occur for reasons related to social behaviour.

Prior to the birth the queen cleans herself thoroughly, particularly her ventrum around the nipples and her anogenital area. The residual saliva may leave olfactory cues that help the newborn kittens to locate the nipples. As each kitten is born, she cleans away the birth membranes, and after the delivery of the placenta, cuts the umbilical cord with her carnassial teeth (Houpt and Wolski, 1982). The cleaning is stimulated by the smell and taste of the amniotic fluid, and tends to be more effective for second and subsequent litters than for the first. Between individual births, the queen may be very restless, pacing around the nest. If a placenta is undelivered while this is occurring, her movements will tend to stretch the umbilical cord, reducing blood flow through it so that when it is severed, it bleeds very little. Both placenta and umbilical cord are eaten; at this point a solitary queen faces several days with little or no food, so she may gain significant nutritional benefit from this. Inexperienced or incompetent mothers may occasionally appear not to distinguish the kitten from its placenta, and will eat the kitten as well (Deag et al., 1988).

After the delivery of the last kitten the queen encircles her litter, usually lying on her side, and encourages them to suckle by nuzzling and licking them. Apart from occasional changes of position, she will remain in contact with the kittens for at least the first 24 h. At this stage the mammary glands are producing colostrum, which is rich in antibodies, before it is replaced by normal milk. For the first 3 to 4 weeks the queen spends up to 70% of her time in the nest, and takes complete care of the kittens. She initiates their feeding bouts by lying down beside them, and grooms them regularly, paying particular attention to their perineal areas to stimulate urination and defecation. She either eats the wastes, or deposits them well away from the nest area. As the kittens become more mobile, they may stray out of the nest; the queen responds to the cries of a displaced kitten by grasping it by the scruff of its neck and carrying it back to the nest. This retrieving activity usually peaks in the third week. The queen is highly responsive to these cries, and will pick up and return with kittens that are not her own if they are in the vicinity of her nest. By this stage she will be spending more time away from the kittens, particularly if she has to forage for her own food, and will announce her return by the 'chirp' call.

From the third week onwards the kittens initiate an increasing propor- tion of the nursing bouts themselves, and soon after this the queen begins to bring dead prey items to the nest. Later, she will introduce them to prey that is still alive, putting them into a situation where they can refine their own prey- catching skills. She will also demonstrate the technique of burying faeces and urine, and the correct places to do so. In feral cats, the bringing of solid food starts the process of weaning, although human intervention often complicates the progress of this stage in litters born to family pets or show queens. From about day 30 onwards, the queen becomes less tolerant of the kittens, and may place herself out of their reach for extended periods of time. When with them, she may discourage them from suckling by adopting one of two postures, the crouch (ventrum in contact with the ground, all four pads supporting the body with the pads in full contact with the ground) and the lie (ventrum in contact with the ground and supporting the weight of the body, the legs partly extended or tucked under the body) (Deag *et al.*, 1988). Once the kittens are mobile they will attempt to ambush the mother away from the nest in order to suckle, and at this stage it often appears as if mother and offspring are engaged in a battle of wills. However, if the kittens are not removed the mother will often continue to demonstrate maternal behaviour towards them even when they are grown up, possibly more so if they are her daughters.

The antagonism between kittens and mother at the time of weaning is entirely understandable from a functional point of view. Provided the mother still has milk, the kittens will benefit from continuing to suckle, rather than test out their predatory abilities, and expose themselves to danger outside the nest. On the other hand lactation is highly costly to the queen, something that is easily forgotten when considering pet cats. The precise costs will depend on the condition of the queen at the time of parturition, and the size of the litter. If the queen has been malnourished during the period of gestation, she will tend to perform less active mothering of her kittens, and will be more aggressive towards them (Martin and Bateson, 1988). Litter sizes vary between one and ten, the median being four to five. In litters greater than four, the birth weight of the kittens declines with number. The queen's milk production increases with litter size, but not in direct proportion, with the result that by 8 weeks of age the kittens in a litter of six are about 25% lighter than those in a litter of two. The mother can lose 6 g of body weight per day, or more with a large litter, and severe weight loss at this stage can also lead to increased aggression towards the kittens, although this may be reduced again if the queen's food is then supplemented. It appears that with larger litters the queen forces the weaning process, because there is a direct correlation between a 'dip' in the rate at which the kittens gain weight and a period of a few days in which the queen does not adopt the normal nursing position. After this time, she will nurse them again, but usually the kittens will initiate this (Deag *et al.*, 1988). The chief factor seems to be the size of the litter, rather than the sex of the kittens; although males grow up to be heavier than females, there is no difference in their weights until after 8 weeks, when the males begin to grow faster. In small litters there is often no discontinuity in weight gain, and in behavioural terms the mothers seem more tolerant of their kittens, the exception being the litter of one, when the mother is more aggressive towards her kitten after the normal weaning time than she would be if there were two. Large litters may cause discomfort to the queen by pushing and biting at her nipples, exacerbated by the competition between littermates, making her less inclined to nurse. In functional terms, the queen will have a greater investment in each kitten if it is part of a small litter as opposed to a large one, and in the former case it may pay her not to risk inducing a metabolic stress that could coincide with exposure to an infection. Certainly kittens from smaller litters tend to go on suckling well beyond 8 weeks of age, although it is not known how much nourishment they obtain from this. There is also some evidence that continued milk intake by kittens once weaning on to solid food has begun may cause digestive upset in some kittens. This is due to changes in the kittens' gut morphology and digestive biochemistry at the time of weaning (Bateson and Bateson, 2002). Suckling behaviour may persist into adulthood; instances have been reported of grandmothers nursing their daughters who themselves are nursing their own kittens.

Development of the Kitten

When it is born the kitten is blind, virtually deaf and completely dependent on its mother. From the time it is 8 weeks old it is capable of an independent existence, even though it may often stay with its mother for a longer period. Such rapid development could only occur if the maturation of the various sensory, motor and control systems were highly coordinated, but the system is not entirely rigid. External events can speed up or slow down individual processes; to take a physiological example, the development of body temperature regulation can be hastened by exposure to cold at 2 weeks of age, but by 4 weeks all kittens have similar thermostatic abilities whether or not they have been exposed to cold. This adaptability suggests that each system has a goal of development, which it can achieve from a variety of starting points, by several routes and by speeding up or slowing down different components, even to the extent that they end up occurring in different orders (Martin and Bateson, 1988). This flexibility is presumably built in to cope with the vagaries of weather, litter size, food availability, behaviour of the mother, and so on, so that a competent adult can emerge despite setbacks along the way. The following account therefore applies only to a 'typical' kitten, although ranges as well as averages will be given where useful, and the remarks made above about the bias towards single mothers and confined litters, made in connection with maternal behaviour, apply equally here.

The gestation period of the cat is about 63 days, and kittens in small litters weigh about 105 g when born. Litters may consist of between one and ten kittens, with a median of four or five, although in the wild the number is likely to be reduced by virus infections, to which the kittens are often highly susceptible. Compared with many other mammals their development is slow, and involves a long period of maternal care. When newly born, the kittens crawl along the mother's abdomen to find a nipple, and alternate between suckling and lying still. Suckling can occupy up to 8 h per day at first, although this declines as the kitten matures. The kittens often purr while suckling, and tread or 'paddle' around the nipples, presumably to stimulate ejection of milk. If they are unable to locate the mother, or if they become isolated, cold or trapped (beneath the mother, for example) they use a 'cry' call to stimulate her attention; it is this cry, rather than their visual appearance, that identifies the kittens to their mother. Harmonically structured squeaks and mew-like vocalizations appear at an early stage of the development of the vocal repertoire (Chaadaeva and Sokolova, 2005).

The queen normally has eight nipples, of which those at the rear are often preferred by the kittens. Individual teat preferences appear quickly and the kittens soon establish a 'teat order' where each kitten predominantly uses one or two particular nipples, even when the mother changes the side she lays on to nurse. Contests between kittens often occur at nipples during suckling, although interestingly there does not seem to be any significant relation between kittens' use of particular nipples and their weight gain or milk intake (i.e. the popular nipples are not more productive; Hudson *et al.*, 2009). The kittens appear to use mainly olfactory cues of their mother to identify their primary nipple when suckling – they will attach to nipples of other early-lactating females (see Fig. 4.1a) but are unable to locate their preferred nipple as efficiently as they would on their mother, even after their eyes have opened (see Fig. 4.1b). Kittens may therefore be born with the innate ability to respond with efficient nipple-searching behaviour to hormonally produced olfactory cues on the mother's ventrum, but soon learn more specific odour cues associated with their own mother and their own particular nipple (Raihani *et al.*, 2009).

For the first 3 weeks the mother initiates all the bouts of suckling, but from then on until the end of the first month the kittens themselves take an increasing role in this, until in the second month of life when it is the kittens who initiate suckling and the mother who begins to avoid their advances. They may even try to suckle while she is standing or moving about, and she may then try to block them, or even become aggressive. At the same stage the kittens become much more exploratory, and also begin to exhibit social behaviour towards each other, both play (discussed below) and other social patterns such as flank-rubbing on the mother.

Social contact with the mother during the first 4 weeks is essential to the normal development of the kittens, which otherwise develop a variety of behavioural, emotional and physical abnormalities. They are unusually fearful of other cats and people, exhibit apparently randomly directed locomotory

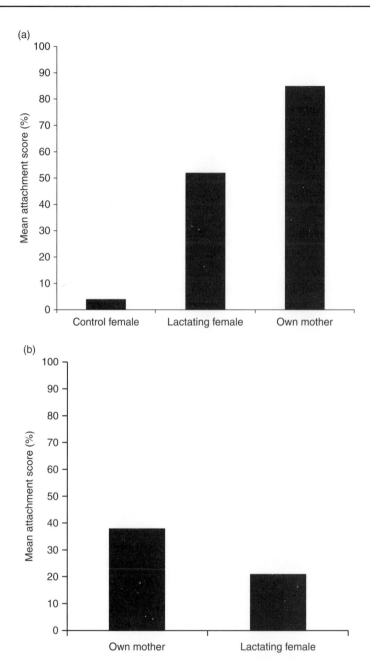

Fig. 4.1. (a) Percentage of kittens attaching to nipples when tested on a non-reproducing female, on a female in the first month of lactation and on their own mother in the first month of lactation. (b) Percentage of nipple attachments by kittens to their preferred nipple when tested on their own mother compared with one in the same location on another lactating female, both in the first month of lactation (redrawn from Raihani *et al.*, 2009).

behaviour and are slow to learn simple associations, such as the location of a source of food. Some of the stimulation provided by the mother can be replaced by simply handling the kitten for 20 min each day, which has an accelerating effect on eye opening, exploratory behaviour and the development of mature electroencephalogram patterns. As has been mentioned above, maternal attention to the kittens can be reduced if the mother is malnourished. If her food intake is 80% of normal, the kittens are buffered from any direct nutritional effects and gain weight as normal, although they do spend more time nuzzling their mother's nipples. More severe deprivation for the mother results in kittens with smaller brains, particularly the cerebrum, cerebellum and brain stem. At 4 months old, male kittens from these litters are more aggressive than usual, and both males and females have motor deficits that cause them to run erratically. The latter at least seem to be primarily affected by the shortage of protein in the diet, rather than the number of calories available (Martin and Bateson, 1988).

General activity does not increase smoothly but proceeds in a series of steps, until it reaches adult levels (Levine *et al.*, 1980). There is a sudden increase between 9 and 14 days, reflecting the maturation of the auditory system, the opening of the eyes and the beginning of walking. A second step occurs during the fourth week, when running is added to the locomotory repertoire, and a third towards the end of the second month, as locomotor play becomes more intense.

The pattern of sleep also changes over the first 2 months. The total time asleep stays roughly constant, at around 60% of each 24 h, until about 6 weeks old when this decreases to about 40%. The type and pattern of sleep, however, change from 100% paradoxical or dreaming sleep (REM) sleep recorded on day 10 after birth to only about 50% REM sleep by day 28. Non-REM or 'quiet' sleep increases reciprocally, and the periods of wakefulness become longer as motor behaviour by the kitten increases (Herman *et al.*, 1991).

Preliminary results from a developmental study looking at behavioural differences among littermates suggest that even during the first month of life individual differences in general behaviours, possibly indicative of later personality, can be identified in kittens. Observer ratings of the behaviour of kittens from three litters were in sufficiently good agreement to produce descriptors of behaviour along the two continua 'Solitary/Contact-seeking' and 'Uninterested/Curious' (Hudson *et al.*, 2010).

Reflexes

At birth, the kitten's behaviour largely consists of simple reflexes (Villablanca and Olmstead, 1979), most of which are directed towards obtaining milk from the mother. Kittens locate the nipple and are stimulated to suckle by a variety of cues (see above); the smell of the mother and her warmth guide them to the ventrum, tactile stimuli detected by their lips guide them on to the nipple, and tactile and chemosensory cues on the tongue stimulate suckling. Nipple attachment is achieved by the rooting reflex, in which the head is pulled back and then lunged forward while the mouth is opened. Once the nipple is located the suckling reflex, which is already present 12 days before normal parturition, takes over.

Neonatal kittens are capable of a certain degree of locomotion, which they achieve by pulling themselves along with their front legs while paddling with their weaker hind legs. During such movement, which may be random or directed towards sources of warmth, the head is often turned from side to side, presumably in an attempt to locate the mother. By the fourth day of life kittens are capable of travelling up to half a metre on their own, orientating themselves by olfactory cues derived from their own body smells. This ability is put to use if they are displaced from their nest – for example, if they have continued to cling on to their mother for a few seconds after she has got up and left. The motor cortex for coordinated forelimb movement matures during the first 2 weeks, and for the hind limbs during the next 2 weeks. By 2 weeks kittens demonstrate immature patterns of forelimb–hind limb coordination, so that rudimentary walking begins at about 3 weeks of age. At the same stage, the reflexes for ground sniffing and the crawl are disappearing. By 4 to 6 weeks two speeds of overground locomotion are used, with walking and running being characterized by different limb coordination patterns, more typical of an adult cat (Howland et al., 1995).

One reflex that can be readily seen is flexor dominance of the vertebral musculature, which is triggered when the kitten is picked up by the 'scruff', the loose skin on the base of the neck. The limbs become limp, the tail is curled and on release the kitten appears to be startled by its new surroundings, as if its sensory systems had been suppressed while it was being carried. It is presumably adaptive for a mother carrying a kitten to assess and react to any danger that occurs, rather than have the kitten react with the concomitant risk of being dropped. This reflex persists into adolescence and can still be triggered in many adult cats (Pozza et al., 2008; see Chapter 3, this volume).

Voluntary urination and defecation appear at about 3 weeks of age, and have completely taken over from mother-stimulated reflex eliminations by 7 weeks. The air-righting reflex appears at about 24 days (range, 21–30), and is perfected by about 40 days (range, 33–48). The orienting reflex is first seen between 8 and 12 days, and begins to be associated with investigatory behaviour from 1 to 4 days later.

The 'gape' response to cat urine appears at about 5 weeks of age, and is fully expressed at 7 weeks. Since kittens have well-developed olfactory abilities considerably before this, it is possible that the gape depends on maturation of the vomeronasal organ. Another stereotyped response to a species-specific signal, piloerection towards the silhouette of a threatening cat (tail-down, arched back, erect ears) is complete by 6 to 8 weeks (Kolb and Nonneman, 1975), although certain components, including the arched back, are present at birth, and piloerection and pupil enlargement appear at about 3 weeks.

The senses

Kittens are born with only four senses functioning efficiently (Fig. 4.2). Tactile sensitivity appears about midway through gestation, and the vestibular balance organ functions even in kittens born a week prematurely. Olfaction is present at birth, and is immediately essential for guiding suckling behaviour. By 3 weeks of age the sense of smell is essentially mature. Taste may be less important, because kittens less than 3 weeks old will suckle from non-lactating females. Hearing and then vision develop after birth.

Hearing

At birth, the auditory system is capable of transducing sounds but the canal is blocked by ridges of skin. These gradually open up as the canal itself widens, over the first 10–15 days. Over the same period the pinna enlarges and becomes corrugated at its base. The pinna can be moved almost from birth, initially both spontaneously and in a diffuse way to a wide range of non-auditory stimuli including tactile, visual and olfactory. By the third or fourth week the pinna movements become more discrete and adult-like. The first responses to sounds come on about day 5, before the auditory canal is open, and consist of general arousal (sniffing, lifting the head and head movements), bouts of pinna movements and squinting of the eyes. Species-specific calls, such as the kitten-call of the nursing mother, the growl or the kitten distress call, elicit responses more effectively and earlier in development than do artificial sounds. By day 16

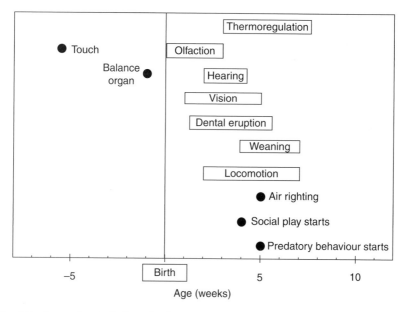

Fig. 4.2. Approximate timings for the appearance and development of selected sensory and behavioural abilities in kittens (from Martin and Bateson, 1988).

directed head movements make it clear that the kitten can locate the source of a sound. By 3 to 4 weeks of age the kitten shows that it can discriminate between different kinds of calls; it tends to approach kitten and mother calls, but withdraws with ears flattened from growls and male threat calls (Olmstead and Villablanca, 1980).

Vision

The kitten's eyes open at between 2 and 16 days after birth. The chief source of this variability seems to be genetic, although the age of the mother and the sex of the kitten also have an effect; female kittens open their eyes earlier than males. At the moment of eye opening vision is barely functional, partly because the optic fluids do not become completely clear until another 4 weeks have elapsed, and partly because the visual cortex requires visual experience before it can mature. Therefore, the acuity of the eyes is initially poor, but improves 16-fold between the second and tenth weeks. Visual cues gradually become important over this period; kittens can usually recognize their mother by her appearance, rather than by her smell and her warmth, by the end of their third week (Martin and Bateson, 1988).

It may seem surprising that a sense as important as vision does not mature spontaneously in the cat. The requirement for visual experience seems to be driven by the cat's binocular vision, because in the rabbit and other mammals that have almost panoramic vision, the visual abilities are only slightly affected by early experience (Blakemore and Van Sluyters, 1975). A certain amount of flexibility seems to be necessary to allow matching of the receptive fields in the two eyes, so that the neural mechanisms for stereoscopic vision and depth perception can be built up. At eye opening, many neurons in the central cortex will respond when either eye is stimulated. If the visual signals from the two eyes are widely disparate – for example, the kitten is born with a severe squint – these cells lose their binocular properties and segregate into two groups, approximately equal in number, each of which responds to only one eye. The exact matching of the images from the two eyes is achieved by the emergence of orientation-selective neurons (see Chapter 2, this volume), which respond to high-contrast edges. In kittens under 3 weeks old, orientation-selective cells are very broad-banded by adult standards, and therefore presumably provide rather poor binocular vision. However, it seems that this initial approximation is enough to stimulate the development of much more specific cells, which appear very rapidly and are essentially adult in character by the time the kitten is 4 or 5 weeks old. Some of these cells respond to spots of contrast, others to vertical contours and yet others to horizontal contours. If any of these elements is completely missing from the kitten's environment (which could only be achieved by deliberate deprivation), the corresponding neurons fail to develop, and it has also been shown that the cat must not only see these features, but react to them behaviourally for the complete set of orientation-detectors to develop. The infantile neurons, now redundant, degenerate, and once the kitten is 3 months old the adult pattern cannot be reversed, nor can it be induced if it has not already appeared. Research into orientation plasticity in kittens continues and now appears to have two phases (see Tanaka *et al.*, 2009 for more detail).

The period from eye opening to 3 months is known as the critical period for vision in the cat. At one time it was thought that the development of many behavioural patterns could only take place in critical periods, but these have proved difficult to demark, particularly for mammals, and have been replaced with concepts of goal-directed development, as mentioned at the beginning of this section.

The initial coordination of vision with movement is largely achieved in the second half of the first month of life (Norton, 1974). Kittens begin to orientate towards visual stimuli about 5 days after they are able to orientate towards sounds (Table 4.1). It is possible that the latter ability only appears when motor development allows but, if so, the former must therefore result from either a maturation of the visual sense or its connection through the visual cortex to the motor cortex. Recognition of the three-dimensional nature of the visual world develops later still, at about 4 weeks of age. Depth perception is often tested using a visual cliff. In this simple apparatus, a sheet of glass is placed over the edge of a step, usually about half a metre high; the horizontal surfaces, both before and after the vertical drop, are painted with a chequerboard pattern. As the kitten approaches the edge of the step, it should be able to see the drop in front of it, but at 24 days old it will almost always continue its progress, supported by the glass surface. Sometime over the next 2 weeks it will begin to hesitate on the 'edge', and turn back. At almost exactly the same age the kitten will also learn to take avoiding action as it blunders about the area of its nest, so it seems that around this age the connection is established between images and the objects that produce them.

Table 4.1. Ages at which various orientated reactions occur to certain auditory and visual stimuli. Descriptions of the visual cliff and visual placing tests are given in the text (adapted from Norton, 1974).

	Age of kitten (days)	
Response	Earliest positive response	Latest negative response
Orientating to a sound	11	16
Following a moving sound	15	21
Orientating to a visual stimulus	16	21
Following a moving visual stimulus	18	24
Avoiding a visual cliff	25	37
Reflex visual placing of forelimbs	27	36
Avoiding obstacles around the nest	26	35
Guided visual placing of forelimbs	33	39

Also at about this time, coordination between the eyes and limbs develops (Table 4.1). This can be simply shown by the visual placing response. If a kitten is lowered gently towards a visible surface it will extend its forelimbs automatically, as it would when landing from a jump. Initially the response is undirected – the kitten sees the surface, and extends its legs down in a reflex manner; if there are holes in the surface, it is just as likely that a paw will go through one of these as land on a solid area. Not for another week or so are the paws actually guided on to the surface by direct coordination between eye and limb (Hein and Held, 1967).

Learning abilities of the kitten

The way that the development of visual abilities depends on actual visual experience is in itself a kind of learning. From the very first days, kittens show that they are also capable of making associations between stimuli in their environment. The first overt indication of this is usually the appearance of nipple preferences, guided by olfactory and textural cues from the mother's ventrum. Using an artificial mother, consisting of a carpeted surface with two rubber nipples, Rosenblatt (1972) has shown that a 2-day-old kitten can learn to distinguish between a nipple that delivers milk and one that does not, based on its texture alone. Discrimination based on odour is possible just 1 day later and is very sensitive. Raihani *et al.* (2009) showed that nipple attachment depends on the female's reproductive state, with virtually no attachments on non-reproducing females, some on pregnant females, the greatest number on early-lactating females and fewer on late-lactating females. Learning also plays a crucial role in orientation back to the nest area. For the first few days, an isolated kitten will crawl in circles wherever it is, but from about day 6 this only occurs in the nest; elsewhere, despite still being blind, kittens can orientate themselves towards the nest using learned olfactory cues, although their initial strategy is usually to cry to attract their mother's attention. Odours from the mother or from littermates are both adequate signals. At 2 weeks of age the same mechanism is still used, but is effective over longer distances, of up to 3 m. From then on the kittens will leave the nest voluntarily for short periods, but it is not until they are 3 or 4 weeks old that visual cues dominate over olfactory ones in guiding them home.

Learning by observation

The mother's influence is not restricted to the provision of food and protection, for in the extended period that the litter stays with her she does a great deal to stimulate their learning. We have already seen that adult cats can learn tasks more quickly if they can observe another, more experienced, adult performing that task, so it would be surprising if kittens did not also learn by observation. It has indeed been shown that kittens can learn to perform an arbitrary task (pressing a lever in response to a flickering light) very quickly if they can watch an adult perform the same task, while kittens that are presented with the task without ever seeing an adult perform it learn very slowly, if at all (Chesler, 1969). Such a large difference in speed of learning might not have been apparent if the task had been based on more ecologically relevant motor patterns, but it does appear that the opportunity to observe adults promotes learning in kittens. In the same series of trials, it was found that mothers are more effective demonstrators than other females. This was particularly marked in terms of the number of sessions taken before the kittens began to succeed (Fig. 4.3); once they had begun to perform correct responses, the time taken to improve to the point of giving the correct response on every occasion was the same whether their mother or another cat was the demonstrator. The difference appeared to lie in how much attention the kitten initially paid to the demonstrator. Since it might be expected that a kitten would initially be apprehensive in the proximity of a stranger, this explains the greater attention paid to the mother in the first few sessions.

The development of predatory behaviour

The most important skills that the mother demonstrates to her kittens are those they will use later for hunting. From the beginning of weaning onwards, she brings recently killed prey to the nest; later, the prey is brought in alive and

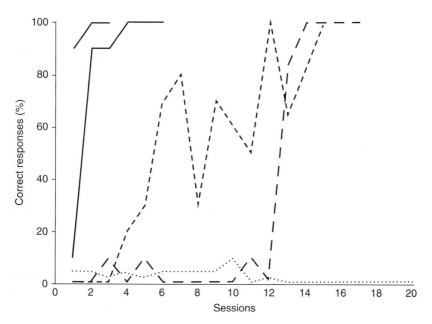

Fig. 4.3. The acquisition of a response (pressing a lever) in representative kittens from three groups given different opportunities to learn. The solid lines indicate the proportion of correct responses in kittens allowed to observe their mother completing the same task; thin and thick broken lines, two kittens that observed a stranger; dotted line, a kitten that was given no opportunity to observe (redrawn from Chesler, 1969).

released near to the kittens. Thus she puts them in a position in which their innate responses, many similar in form to those used in play, bring them into contact with prey at an age at which they would find it impossible to achieve the same degree of contact on their own. Thus the goal of hunting and the link between predatory behaviour and food are firmly established at an early age. The role of the mother in this is part demonstrator and part supervisor, in that she sets up the situation in which the kittens can learn, but does not necessarily perform the required patterns in front of them. However, if the kittens hold back from attacking the prey, she is likely to initiate an attack herself (Caro, 1980).

Kittens' tendency to follow their mother's lead extends to food choice as well. They imitate their mother's food selections, so that if the mother has previously been trained to eat an unusual food, such as banana or mashed potato, they will also prefer this to more 'natural' foods. The peak of imitative learning of food choice comes when the kittens are 7 or 8 weeks old, and after this their preference for whatever their mother eats persists even when she is not present. A similar process probably takes place with natural prey, because kittens prefer to kill strains of rat that they have seen their mother kill over other strains. This may be an effect largely confined to large prey with the capacity to fight back and cause injury to a young, inexperienced cat – Adamec et al. (1980) found that the observation of prey killing had much less effect on subsequent skill if the prey was a mouse as opposed to a rat. Indeed, a single experience of killing and catching a mouse can turn a kitten into an accomplished mouse killer. The main effect of observation in the case of large prey may be to overcome the fear that conflicts with the predatory drive in inexperienced cats and causes them to react defensively towards rats.

The actual behaviour patterns that kittens use to interact with prey have been classified as 'predatory play', but since such sessions would, in the wild, almost inevitably end with the consumption of the prey, this may be stretching the definition of play (see next section). However, virtually all of the motor patterns seen in object play also form part of prey handling, including poke/bat, scoop, grasp, bite/mouth, belly-up, stand up and pounce, so the link between the two sorts of activity is evidently a strong one. Skill in prey handling during the weaning period (1–3 months) accurately predicts predatory abilities at 6 months and early weaning accelerates the onset of actual killing of live prey, but the origins of the skills themselves are less easy to pinpoint. This is probably because multiple factors influence the development of hunting ability (Fig. 4.4), and indeed one would expect that, for such an essential function as obtaining food, development would need to be highly flexible to cope with the varied ecological situations that different kittens may inhabit.

Play Behaviour

Play is easily recognized, but difficult to define (Martin, 1984). Some of the behaviour of young animals brings about immediate consequences that can

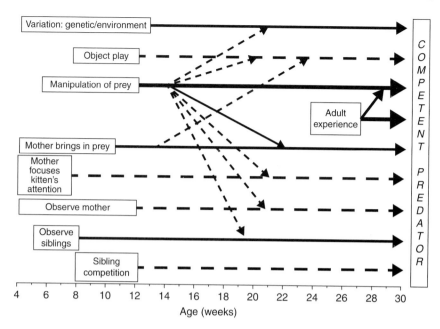

Fig. 4.4. Some of the factors that are thought to influence the development of predatory skills in adults. Proven influences are shown by solid arrows and hypothetical influences by broken arrows. Diagonal arrows indicate some of the possible interactions between influences. The approximate duration of each influence is indicated by the position of its text box.

reasonably be called adaptive – suckling is an obvious example. It is not so easy to find an immediate benefit for many other activities, such as manipulating inanimate objects, chasing littermates or climbing on to obstacles around the nest (although attempts have been made to do so), and such activities are often placed in the category of play. One formal definition is 'all motor activity performed postnatally that appears to be purposeless, in which motor patterns from other contexts may often be used in modified forms and altered temporal sequencing'. Obtaining hard information on the benefits that playing as a kitten brings to the adult cat has proved difficult, and so for practical purposes play is often defined in terms of the motor patterns involved, which are often similar to species-specific behaviour patterns performed by adults, although the context, intensity and sequencing are all likely to be altered when they form part of play.

Perhaps the most convenient way to classify play behaviours is in terms of the object towards which the kitten directs such motor patterns. Social play involves a conspecific, which is likely to be a littermate but can also be the mother or other adult. Play can also be focused on inanimate objects – these are treated differently depending on whether they are large and immobile, in which case they are climbed on and jumped off; or small and mobile – in which case they appear to become substitutes for prey items. The latter type of object play can also be seen directed at invisible objects – for example, kittens may pounce on 'thin air'. It is impossible to tell in such cases whether the kitten really has imagined a prey item or whether a speck of dust has triggered a set of behaviour patterns normally reserved for larger objects. Kittens also chase their own tails, such self-directed play forming a category that has received little attention from cat behaviourists. However, locomotor play, object play and social play have all been studied in detail in the cat, and each will be described before their possible functions are discussed.

Locomotor play

Adult cats are extremely agile, and those parts of the nervous system concerned with balance and locomotion are correspondingly well developed. Some of the skills that contribute to agility appear almost fully formed – for example the air-righting reflex, whereas others develop more slowly and seem to improve with practice. Once they have left the nest, kittens explore their environment vigorously, not only by looking, sniffing and listening, but also by clambering over any objects in their path, including those that are unstable or will not bear their weight; gradually they build up experience of which types of objects are safe to walk or climb on, and which are not. Much of this exploratory locomotion appears to have little short-term benefit (it may even put the kittens at risk by making them conspicuous to predators) and so may be classified as play; certainly it appears to be rewarding to the kitten for its own sake.

This aspect of play has not received the same attention as object play and social play, but is nevertheless a crucial aspect of kittens' development. A study by Martin and Bateson (1985) described many of the important features of locomotor play. Seven litters of kittens were given access to a wooden climbing frame every few days, from the time they were 36 days old to the end of their second month. To begin with they spent only about 3 min of each 30 min test period on the frame, but by 7 weeks old this had increased to almost 20 min. For comparison, their mothers spent between 1 and 5 min on the frame during the same periods. The frame was used for social as well as locomotor play, but the latter predominated, possibly because opportunities for social play were also available outside the test sessions. Until they were about 7 weeks old, the kittens largely restricted their activities to the lower rungs of the frame, presumably because their locomotor systems were not sufficiently mature until then. Some had still not ventured to the highest parts by the end of their second month, and those that did tended to lose their balance more often than those that did not, although such slips rarely resulted in any kitten actually falling off the frame. After these occasional falls, the kitten concerned would normally climb straight back on to the frame, confirming the self-rewarding nature of such play. Male kittens were no more or less exploratory than females, but there were considerable differences between families. Some of these differences appeared to be linked to the mother's behaviour; kittens that eventually made use of the top of the frame were those whose mothers had spent the most time on the frame during the first few sessions.

Object play

Confronted with novel objects, kittens will investigate them carefully by looking, sniffing, licking and touching, and will often circle around the object to repeat this process from a number of angles. Subsequently, small objects are usually played with by being poked, batted, grasped and tossed in the air. Such bouts of play are often preceded by pounces or leaps reminiscent of the predatory behaviour of adults. Larger objects can be used as hiding places or perches, and for locomotor play.

Descriptions of some common object play patterns are included in Table 4.2. Most of these are not performed in isolation, but form sequences in which the various patterns occur in predictable, though not fixed, orders. Bouts of object play tend to go on for much longer in felids than they do in the young of other carnivores, usually because similar sequences are repeated over and over again. The poke/bat pattern may be crucial to stimulating this repetition, because it makes the object move, whereupon it can be chased and pounced on (Aldis, 1975). In one study, 35% of play bouts were started by poke/bat, which also made up 29% of patterns observed within sequences (West, 1979). Poke/bat was also often followed by another poke/bat, with as many as 11 repetitions being recorded. The other common initiating pattern was vertical stance (21%), which also made up 15% of behaviours within sequences. Other common components were bite/mouth (12%), chase (10%) and belly-up (9%). The scoop pattern is less common than any of these, but has an interesting variation, seemingly preferred by cats but by few other carnivores, in which it is performed through a hole at an object on the other side. In general, although object play has been compared to predatory behaviour, some of the sequences are also reminiscent of the adult cat's post-hunting manipulation of prey, which will be discussed further in Chapter 7.

Social play

Kittens begin to initiate play with their littermates as soon as their sensory and motor systems are mature enough, and continue to interact at a high rate for the next few weeks. Many of the behaviour patterns seen are similar to those used in object play (Table 4.2), modified slightly to take account of the liveliness of the target, while others are only seen as components of social play. The latter – for example the arched back component of side-step and horizontal leap, are similar to patterns of social interaction seen in older cats, whereas the former are more predatory in form. West (1979) described the most common patterns for initiating social play as pounce (39%), belly-up (14%) and stand up

Table 4.2. Some definitions of behaviour patterns seen in kitten play (from West, 1979). Not all authors have used precisely these categories – for example, similar but not identical descriptions can be found in Barrett and Bateson (1978) and Caro (1981). The figures in parentheses indicate the time when each social pattern first appears (according to West, 1974).

Poke/bat	Kitten contacts an object with either of the front paws from either a vertical (poke) or a horizontal (bat) orientation (Fig. 4.5)
Scoop	Kitten picks up an object with one of its front paws, by curving the paw under the object and grasping it with its claws
Toss	Kitten releases an object from the mouth or paw with a sideways shake of the head or paw
Grasp	Kitten holds an object between the front paws or in the mouth
Bite/mouth	Kitten places an object in the mouth and closes and opens its mouth around the object
Belly-up	Kitten lies on its back, belly up, with all four limbs held in a semi-vertical position; the hind limbs may move in treading movements and the forelimbs may be used to paw at an object or kitten; the mouth is typically open (22 days) (Fig. 4.6)
Stand up	Kitten stands over an object or another kitten (the target); if the latter, its head is oriented towards the head and neck region of the other kitten; the kitten's mouth is open and it may lunge with its mouth, or raise a paw, towards the target (24 days) (Fig. 4.6)
Vertical stance	Kitten extends its hind limbs so it is in a bipedal position with the forelimbs outstretched (36 days)
Pounce	Kitten crouches with its head touching the ground or held low, its hind limbs tucked in and tail straight back, often moved back and forth; kitten then treads with its hindquarters before leaping forwards and upwards by rapidly extending its hind legs (34 days) (Fig. 4.7)
Chase	Kitten runs after a moving object or kitten (39 days)
Side-step	Kitten arches its back, curls its tail upwards and towards its body, and walks sideways (33 days) (Fig. 4.8)
Horizontal leap	Kitten arches its back, curls its tail upwards and towards its body, and leaps off the ground (43 days)
Face-off	Kitten sits near another kitten and hunches its body forward, moving its tail back and forth and lifts a paw towards the other, often with a swiping motion (45 days)

(16%). These were generally about 90% effective in eliciting a response from the other kitten, and were used in the same proportions at both 6 and 12 weeks of age. Two patterns that changed in frequency of use were vertical stance (8% at 6 weeks, 24% at 12 weeks) and side-step, which declined from 20% to 3% over the same period. Side-step was generally only about 75% effective in

Fig. 4.5. Poke/bat (see Table 4.2).

Fig. 4.6. Two postures seen in social play – stand up (right) and belly-up (left) (see Table 4.2).

Fig. 4.7. The starting posture for pounce (see Table 4.2).

obtaining a response, so it is possible that between weeks 6 and 12 the kittens learned to avoid using this behaviour pattern at the beginning of sequences, because other initiations were more likely to achieve the desired bout of play. Another unusual attribute of side-step was that it tended to elicit a side-step in the responding kitten, whereupon the initiator kitten would perform another side-step, and so on. Other common patterns tended to elicit a complementary, rather than identical, response. So, for example, pounce was often reciprocated by belly-up, belly-up by stand up, stand up by belly-up, and vertical stance by belly-up. Such alternations tend to bring about a reversal of roles within the pair of kittens as the interaction proceeds, each taking it in turns to be initiator and recipient. Bouts were most often terminated by two specific patterns, chase and vertical leap.

Fig. 4.8. Side-step (see Table 4.2).

The goal of such social play appears to be physical contact between the kittens. Since several of the typical behaviour patterns are similar to those used by adult cats during fights, it is worth considering why bouts of social play do not escalate into truly agonistic encounters. One simple answer is that they occasionally do, and this becomes more common as the kittens get older and as elements of sexual behaviour become incorporated into the repertoire of male kittens. However, the majority of bouts are apparently amicable, and the observation that each kitten alternates between being the 'aggressor' and the target indicates that there is little, if any, true aggression involved. It has been suggested that certain signals indicate the intention to play rather than to fight. One is the half-open mouth, the kitten's 'play-face', often displayed at the beginning of the interaction, and during such patterns as belly-up. Another is the ritualized nature of many of the patterns, always performed at the same level of intensity and in a stereotyped way, which should tend to enhance the confidence of the recipient. Signals produced by moving the tail at particular rates have also been suggested as indicators of an intention to play. What appears to be 'play-fighting' may also take place between adult cats, sometimes escalating into genuine aggression and sometimes not. The factors that induce or inhibit such escalation have not been studied, but are very likely to include what each individual predicts the other might do, through their previous experience of each other.

All of the above presupposes that there are at least two kittens in the litter; occasionally only one is born, or disease leaves a single survivor. Social play in one-kitten litters has been studied by Mendl (1988) – one interesting finding was that single kittens did not engage in any more object play or self play than did kittens in litters of two. This indicates that, despite the overlap in motor patterns between object and social play, they are separately motivated, a conclusion also supported by the ages at which each type of play is most intense (see below). The solitary kitten must direct all its social play towards its mother, and since the mother is much more likely to be absent from the nest than is a sibling, it is not surprising that single kittens are involved in relatively little social play. However, they are able to compensate partly for what appears to be the preferred type of social play, with siblings, by initiating bouts of play with the mother, and the mother does respond with play behaviour patterns herself. This continues up to the time when the kitten is about 2 months old, when she begins to respond aggressively towards invitations to play. There is no evidence that the mother somehow 'realizes' that she should play with a single kitten, because otherwise it will receive none; it is the kitten, not the mother, that initiates each play bout, and the mother may simply be responding to signals that would be directed almost exclusively at siblings in a larger litter. Also, the types of behaviour that the solitary kitten pursues are altered; it can usually stalk its mother and paw her tail without disturbing her unduly, but she seems unwilling to reciprocate the more intense patterns such as wrestling and chasing. If, as has been suggested, activities involving body contact are the goal of social play, then the quality of play for the single kitten may be reduced. The long-term consequences of this for the kitten – in particular, its social behaviour as an adult – are unknown.

The development of play

The various components of play are not expressed in the same proportions as the kitten grows. Moreover, social play and object play, while overlapping considerably, peak at different stages of development. The precise timing of all these changes can be affected by genetic factors, the behaviour of each individual mother and by environmental variables such as the availability of food. The typical timings given below should therefore be taken as approximate, and as applying directly only to kittens housed indoors with their mothers.

Many components of social play are first seen as isolated, self-directed patterns, often repeated over and over again as if being practised – the approximate times at which each appears are indicated in Table 4.2. In the fifth or sixth week mixed sequences appear, and are directed towards other kittens for the first time as they start to hide when moving towards another kitten. At this time they will also start searching for an object that has disappeared. During week 7 these behaviours will combine into playful social interactions in the form of hide and seek (Dumas and Dore, 1991; Bateson, 2000). Complex sequences develop rapidly, but the vertical stance and face-off patterns are not common components of sequences until several more weeks have elapsed. During the first 3 weeks or so of social play, three or more kittens are often involved in the same sequence, but by about the ninth week the great majority only involve a pair. The peak of social play is a broad one, from about the ninth to the 14th weeks.

Object play occupies a smaller proportion of the kittens' time (although it may occur quite frequently, but briefly, during bouts of social play) until about week 7, when it increases in frequency and can become commoner than social play by about week 16, after which its frequency seems to depend considerably upon the 'character' of the kitten and the conditions under which it finds itself. This changeover occurs in parallel to changes in other types of social behaviour. For example, kittens approach one another more, touch noses more and are more likely to make contact with one another in weeks 10–14 than in weeks 18–21, but in the latter period they are more likely to be genuinely aggressive towards one another (as distinct from the 'mock' aggression of social play; Mendoza and Ramirez, 1987). There appears to be a major change in the organization of play activity in the fourth month, which correlates with, but is not necessarily driven by, changes in the environment in which the kittens find themselves. From birth until the fourth month, the kittens are confined to a small area, because at least initially their motor capabilities are insufficient to take them far from the nest, and in any case they need to be close to their mother for nourishment. Siblings are therefore readily available, and opportunities for social play are all around. The fourth month is the time that feral cat mothers often leave their kittens, and coincident with this there is an increased tendency to explore the environment, and to play in, on and with objects, even in kittens that are still housed with their mothers.

There also appears to be a more subtle change in the quality of play at the end of the second month, about the time that weaning is normally completed. Comparing different litters, Caro (1981) found that some patterns, such as pounce, poke/bat and bite/mouth (using the terms in Table 4.2), could be grouped together, and showed a positive association with the components of the emerging predatory behaviour, while others, including the arched back of horizontal leap and side-step, vertical stance and chase, became progressively dissociated from predatory behaviour patterns. This may indicate that, while a single motivational system may control social play when it first appears, as the kitten develops certain components may be elicited differentially by two systems, one connected with predatory behaviour and the other with agonistic social behaviour. There are, of course, other explanations for some of these changes; a possible reason for the decline in the frequency of vertical stance, based on learning, has already been mentioned.

Several studies have examined variations in the process of weaning on the expression of play. In particular, the switch from social to object play seems to follow the completion of weaning, and might therefore be driven by this event. If the mother cannot obtain as much food as her normal *ad libitum* intake, she will tend to wean her kittens earlier than usual, even if their growth is unaffected. This stimulates both object and social play in the kittens. The mother appears to achieve this directly, and at a much earlier stage than when object play itself appears, by reducing some aspects of maternal care during the first 3 weeks after the birth (Bateson *et al.*, 1990). While one might predict that inadequate nutrition should lead to a reduction in a 'useless' activity such as play, in fact play is not energetically expensive (see below) and increased object play may be a way for the kittens to gain predatory skills more rapidly if it seems likely that food is going to be hard to find.

Sex differences in play behaviour

Until about 19 weeks of age there appear to be few qualitative differences in social play between male and female kittens. At that time the sexual behaviour of males begins to appear, and they will sniff the females' anogenital areas, attempt the neck grasp and try to mount. Females often respond aggressively to such advances until they begin to reach sexual maturity, from about 23 weeks. Sex differences in object play emerge much earlier, from about week 7, when the rate of object play by male kittens begins to rise to about twice the rate performed by females (Barrett and Bateson, 1978). This is not due to a simple effect of motor development, since at this stage females are actually more active than males. The increase in object play is particularly marked in single-sex litters, since in mixed litters the females tend to play almost at the same intensity as males. A similar 'masculinization' effect is also detectable in social play between 12 and 16 weeks, such that all-female litters play slightly less aggressively than do mixed or all-male litters.

The functions of play

Although there has been a great deal of speculation about the functions of play in cats, and in many other mammals, little hard evidence has emerged supporting any universal theory (see Held and Spinka, 2011). The underlying notion has been that the play of the young animal somehow produces benefits later in life, although this has been challenged by some authors who have speculated on benefits that the young animal might gain almost immediately and which might be undetectable in the adult. The search for adaptive benefits has also tended to assume that play involves the young animal in considerable costs that must be recouped later in life, otherwise kittens that did not play would tend to survive to adulthood, breed and perpetuate whatever genes were available to suppress play. In fact, it has been shown that the metabolic costs of playing are slight – a kitten that plays for a typical 9% of the day will increase its energy expenditure by an average 4%. There may be other costs to free-living cats, such as increased exposure to predators during locomotor play, but these have not been quantified.

The presumed functions of play can be divided into three categories – motor training, cognitive training and socialization. Much of the research on cats has focused on the former, and in particular on the impact of play on subsequent predatory skills. Comparison of the cat with the dog shows that many differences in their play behaviour can be directly correlated with differences in their predatory behaviour. For example, puppies indulge in violent bouts of shaking of objects held in their mouths, reminiscent of the killing shake of wolves, whereas cats rarely do so. Object play and hunting are both solitary activities for cats, but for dogs these are social and cooperative activities, respectively. Stalking and ambushing are important parts of both social and object play in cats, but dogs rarely use these in hunting or perform them during play (Aldis, 1975). There are at least two possible interpretations of these similarities, however. One is that performance of predatory motor patterns during play improves the motor skills of the young animal, so that it becomes a more efficient predator later in life, and it is therefore logical that practise of the predatory motor patterns themselves is the best preparation for the real thing. Another explanation for the similarity between play and predation is simply that the predatory motor patterns are already to some extent programmed into the motor, sensory and control systems of the young animal, and it would be inefficient for a complete set of play-specific patterns to be programmed as well, particularly as they would, by definition, only be active for a few weeks in the whole lifetime of the animal. Thus, if possible, play should 'borrow' motor patterns from those used universally by adults. Following this line of reasoning further, even if the primary function of play were not the practise of predatory skills, we might expect to see predatory patterns used because they are already in the cat's repertoire and, with some modification, could be used to facilitate cat–cat interactions.

Certainly it has proved very difficult to determine any firm link between the degree of object play or social play in kittens, and subsequent predatory skills in those same individuals. This is probably because there are all kinds of factors that could, or have been shown to, influence the competence of adult hunting, and none is in itself essential (Fig. 4.4). Since kittens in the wild are at great risk immediately after weaning, it is possible that it is only at this stage that play has a significant effect on hunting success; by the time the kittens that have survived have reached adulthood, those that were less competent at weaning have caught up and so no effect of play is then detectable (Martin, 1984). This fits in with the apparently goal-directed nature of much kitten development, as discussed earlier, and with the observation that early weaning stimulates object play.

Less attention has been paid to the possible role of play in forging cognitive skills that will be used later as components of social behaviour. This is unsurprising in view of the comparatively recent discovery of the complexities of adult cat sociality. However, the separation of social play into aggressive and predatory components, probably under the control of separate motivational systems, suggests in itself that this type of play may have functions other than the honing of predatory skills.

Finally, West (1979) suggests that we may be looking too far into the future to discover one function of social play; it may simply serve to keep the litter together at a time when each kitten needs to be easily found by its mother for suckling, and when an isolated kitten may present an easy target to a predator – play as an 'invisible play-pen'.

Cat–Cat Socialization

Learning of species and individual identity has been widely investigated in vertebrates, and usually involves one or more 'sensitive periods' early in the animal's life in which such learning occurs much more readily than at other times in the lifespan (Bateson, 1979).

Studies of the ontogeny of social behaviour in the domestic cat have largely focused on the development of (interspecific) social attachments to humans (see Chapter 9, this volume), and little attention has been paid to the timing or mechanisms of the development of intraspecific social behaviour. There is no evidence to indicate that socialization to humans interferes with the development of intraspecific social behaviour – kittens raised with both conspecifics and humans subsequently direct their sexual behaviour only towards their own species. It is therefore unclear whether socialization to conspecifics in this species is truly homologous with interspecific socialization, or merely analogous.

Intraspecific socialization in the domestic cat presumably evolved to establish social bonds with the mother and littermates, and to direct subsequent sexual preferences, since the ancestral species is solitary and territorial. Cats vary substantially in their tolerance of conspecifics, suggesting that intraspecific sociability is a stable and variable characteristic of individuals (Mertens and Schär, 1988). Possible underlying genetic influences may include the 'boldness' trait that affects socialization to humans (McCune, 1995); ontogeny is a more likely explanation for the greater tolerance of littermates for close confinement with each other compared with cohabiting but unrelated cats (Bradshaw and Hall, 1999) and the familial associations between juvenile feral males noted by Macdonald et al. (2000).

Little is known about the sensitive period for intraspecific socialization in this species, although it has been stated that 'social relationships…are most readily formed in the first two months after birth in domestic cats' (Bateson, 2000). The sensitive period may begin at about 2 weeks of age, as soon as the relevant sensory systems become competent, and may persist for the next 6–7 weeks, running in parallel to socialization to people. In one study (Schneirla and Rosenblatt, 1961), kittens removed from their mother between 23 and 44 days of age were most likely to hiss at their mother and never suckle from her; kittens isolated from 34 to 49 days approached the mother and restarted suckling. This could indicate that kittens need to be with their mother for at least half of the period between 3 and 6 weeks of age in order to maintain their bond to her. However, this study addressed only behaviour directed at the mother, not at siblings or other conspecifics. Guyot *et al.* (1980) suggested that interaction with littermates was more important in the development of cooperative social interactions than interaction with the mother. Social play usually declines after 16 weeks of age, suggesting that a sensitive period for the formation of relationships with individuals other than the mother may be ending at about this age.

Hand Rearing

It is possible to hand rear abandoned kittens, although their chances of survival are increased the longer they are able to feed from their mothers. The potential problems of socialization of such kittens do not seem to have been studied in detail, although there are profound effects on sexual behaviour that severely reduce their chances of successful reproduction when adult (Mellen, 1988). While this is of considerable importance to those interested in using the domestic cat as a model for breeding other small felids in zoos, it is unlikely to matter in a shelter where neutering before or soon after homing is the general rule. Hand raising may, however, also affect species identity and produce an animal that is difficult to socialize to other cats; for such an individual, living with conspecifics may actually be stressful. Anecdotally, hand rearing is believed to predispose kittens to showing attention-seeking behaviours because of the high value they place on human social contact; in a small study, J.W.S. Bradshaw and S. Cook (unpublished) found evidence for this in such kittens' behaviour while being handled, but this difference from the behaviour of typical mother-reared kittens did not appear to persist into adulthood.

Changes in Old Age

Adult life is not a behavioural plateau that follows the steep upward slope of learning during kittenhood. As the cat ages, changes occur in the central nervous system that are almost certainly reflected in behaviour. Conditioned reflexes (see Chapter 3, this volume) are harder to establish in older cats, but cats between 11 and 16 years of age seem to be at least as mentally flexible as younger animals – for example, in learning changes in the location of food. McCune *et al.* (2008) found no significant age-related decline in cognitive function when testing cats on a hole-board task, and Levine *et al.* (1980) showed that aged cats were actually superior to younger ones in a reversal learning task. They felt this probably reflected the rapid forgetting of previous learning in aged cats. Subsequent researchers have suggested that neuropsychological testing currently being developed for use on cats may be a better way to investigate age-related cognitive changes – preliminary data from these suggest that senior cats are in fact impaired relative to normal adults (Landsberg *et al.*, 2010). The aim of this research is to elucidate more about the neurodegenerative disorder known as cognitive dysfunction syndrome, which is accompanied by a number of cognitive and behavioural changes in the cat (see Chapter 12, this volume).

If they have remained healthy, cats' locomotory skills remain unimpaired as they enter their second decade, as are their reactions to the vocalizations of other cats (Levine *et al.*, 1980). However, they habituate much less quickly to repeated stimuli than do younger cats. It is known that certain areas of the brain are susceptible to ageing while others are much more resistant, and so selective changes in ability might be expected as cats age. It seems likely that a cat cannot be said to be functionally old until at least its 16th year, although an increased risk of many diseases as age advances may influence behaviour (see Chapter 12, this volume).

Communication

5

Introduction

For many years the communicative skills of domestic cats were undervalued, because it was thought that the average adult cat might need to communicate only the boundaries of its territory and its willingness or unwillingness to participate in aggressive or sexual encounters. Further studies revealed that cats actually have a fairly varied repertoire of social behaviour and can occupy overlapping areas – current knowledge of the mechanics of these social interactions is discussed in Chapter 8. The present systems of communication between domestic cats have arisen as a product of evolution from ancestral Carnivora, and the requirements of the habitat in which *Felis s. lybica* developed. There may have been further changes as a result of domestication, although a study by Mellen (1993) comparing 20 different species from seven lineages of small cat (including the domestic cat) showed remarkable similarity in both scent marking and ordering of social behaviours. Certainly the stability of social groups of farm cats argues in favour of the retention of a full repertoire of ancestral communication patterns.

Particular modes of communication suit different sets of circumstances; before each is discussed in detail it is worth outlining the particular advantages and disadvantages of communicating by sight, sound or smell. Olfactory communication is particularly important in solitary species that do not know when the message they wish to convey will be received. Odours can be active for long periods of time if they have been appropriately formulated, and are placed where they will not be destroyed by the effects of weather. Such signals cannot, however, be turned off at will, which may not present many problems for a

© J.W.S. Bradshaw, R.A. Casey and S.L. Brown 2012. *The Behaviour of the Domestic Cat*, 2nd Edn (J.W.S. Bradshaw *et al.*)

large animal with few enemies, but may be a distinct disadvantage for a prey species. Orientation to chemical signals is at best risky, because over distances of more than a few centimetres they are carried by the wind, and when solitary cats need to locate one another – for example, when a female is in oestrus – sounds are used as well as odours. Chemical signals are most effective when used at night, or underground, or in dense vegetation, circumstances in which visual communication is difficult. Auditory signals are as effective at night as during the day, and depending on their frequency can be made easy or difficult to locate. They are also effective over very long distances, one example from another carnivore being the wolf's howl. Over short distances during daylight almost any medium can be effective, and when unambiguous communication is critical to survival, such as during agonistic encounters, several signals are used simultaneously. Over long distances each modality has distinct properties, some of which are listed in Table 5.1.

Acoustic Communication

Compared with most other members of the Carnivora, pet cats are unusually vocal. Feral cats are generally much more silent, implying that individual cats learn to use vocal communication to varying extents, depending on the environment in which they grow up. More speculatively, it is possible that the ability to learn how to use vocalizations to communicate with humans is itself a consequence of domestication.

The vocal repertoire of the adult cat

A precise figure for the number of distinct sounds produced by cats depends on the authority consulted, and a definitive description of the complete repertoire has been hampered by their huge variation. Authors have tried to classify the range of calls produced in different ways using acoustic, articulatory, phonetic or behavioural criteria (see Nicastro, 2004). There could be at least two reasons why they are so hard to categorize. The first is that while each call may have a 'typical' form, it will not be used on all occasions by all individuals, and a divergent form of one call may sound much like a variation on another.

Table 5.1. Properties of the three main signalling channels when used over long distances.

	Maximum range	Direction	Sensitivity (signal/noise)	Source location	Signal off
Sight	Medium	Straight	Medium	Direct	Rapid
Sound	Long	Omnidirectional	High	Indirect	Rapid
Odour	Long	Of wind	Very high	Very indirect	Slow

Alternatively, some types of call – one obvious example being the miaow – may form a continuum that the various classification systems have divided up to different extents. Furthermore, many of the descriptions of calls come from domestic pets, in which the form of some calls may have been shaped or trained as part of the cat–owner relationship. For example, of the 16 patterns characterized phonetically by Mildred Moelk (1944), three were noted only from her own cat.

Many oriental breeds are self-evidently more vocal than the occidental varieties, some more than others, although no systematic description seems to have been made of their actual repertoire of sounds. Marchei *et al.* (2009), comparing breed differences in the development of kittens, found that their sample of Oriental/Siamese/Abyssinian kittens always vocalized more than their group of Norwegian Forest kittens (a breed known to be slower in development than the former group).

Twelve of the most commonly distinguished domestic cat calls are given in Table 5.2, divided according to the position of the mouth when the sound is emitted. Two, the purr and the trill, can be produced with the mouth closed. Four are produced while the mouth is opened and closed, much as human speech is; these are the miaow, the female and male mating calls (which can also fall into the third category, depending on their intensity) and the aggressive howl. The third group are all 'strained-intensity' calls, produced in a stereotyped way with the mouth held open in a fixed position, and often accompanied by the appearance of great

Table 5.2. Characteristics of vocal calls used by adult cats and the circumstances in which each is most typically used (from Moelk, 1944; Brown *et al.*, 1978; Kiley-Worthington, 1984; and Houpt, 2011).

Name	Typical duration (s)	Fundamental pitch (Hz)	Pitch change	Circumstances
Sounds produced with mouth closed				
Purr	2–700	25–30	–	Contact
Trill/chirrup/ murmur	0.4–0.7	250–800	Slight rise	Greeting, kitten contact
Sounds produced while the mouth is opened and gradually closed				
Miaow	0.5–1.0	700–800	–	Greeting
Female call	0.5–1.5	?	Variable	Sexual
Mowl (male call)	1.0–2.0	?	Variable	Sexual
Howl	0.8–1.5	700	–	Aggressive
Sounds produced while the mouth is held open in one position				
Growl	0.5–10	100–225	–	Aggressive
Yowl	3–15	200–600	Rising	Aggressive
Snarl	0.5–0.8	225–250	–	Aggressive
Hiss	0.6–1.0	Atonal	–	Defensive
Spit	0.01–0.02	Atonal	–	Defensive
Pain shriek	1.0–2.5	900	Slight rise	Fear/pain

tension in the facial muscles. Most, apart from the version of the female call used during courtship, are associated with aggressive or defensive behaviour.

Purring

The purr is highly characteristic of the felids, but the way that it is produced has remained a mystery for many years and its biological function is still not fully understood. It is easy to hear the purring that stroking often elicits in a pet cat, and it is this sound, along with various miaows, that pets most often direct towards their owners. However, by fitting free-ranging cats with throat microphones, Kiley-Worthington (1984) found that purring occurs in a much wider range of circumstances than just cat–human interactions. Purring was not only switched on by the presence of the cat's handler, but also while the cat was nursing kittens or greeting a familiar social partner, during tactile stimulation such as rolling or rubbing, drowsy sleep and by warm, familiar environments. Conversely, a variety of strong stimuli switched off purring, including aggressive or sexual interactions, a social interaction with a cat not encountered for a long time, and while hunting and in the presence of prey. It is difficult to summarize the various circumstances under which purring is heard, but all appear to be associated with actual cat–cat or cat–human contact, or circumstances under which cat–cat contact might be desired, such as when the cat is about to sleep or when performed by suckling kittens; the latter may be a cue inducing the mother to lie still (see below). This idea has been summarized as 'contact' in Table 5.2. Perhaps its most bizarre occurrence is when a cat purrs when apparently in extreme pain – this may indicate that purring retains a care-soliciting function in adult cats as well as in kittens.

There has been a considerable amount of speculation about the way that the purr sound is produced, not least because it is produced continuously, during both inhalation and exhalation, except for a brief pause at the transition between the phases of the respiration cycle. It is now known to be a true vocalization, because it involves the vocal apparatus. The sound is due to a sudden build-up and release of pressure as the glottis is closed and then opened, and the sudden separation of the vocal folds as a result of the pressure build-up. The glottis is moved by laryngeal muscles that contract for 10–15 ms every 30–40 ms. These are driven in turn by a very precise, high-frequency neural oscillator that appears to be free-running, since no reflex arcs to mechanoreceptors in the region of the vocal apparatus can be detected (Frazer Sissom *et al.*, 1991).

It is now thought that purrs can be combined with other vocalizations to change their meaning. Purrs recorded when the cat is seeking food contain, within the low-pitched purring sound, a high-frequency, voice-like component, more like a cry or miaow. Humans can tell the difference between this and a normal, contented purring and rate the former as more urgent and less pleasant than the latter (McComb *et al.*, 2009). This suggests that the food-solicitation purr may have evolved particular acoustic qualities that make it difficult to ignore – a 'manipulative' and care-soliciting signal (Bradshaw and Cameron Beaumont, 2000), which is probably reinforced by owner response.

Miaowing

These vocalizations are rarely heard during adult cat–cat interactions (Brown, 1993) and yet are very common in cat–human interactions, suggesting that cats have learned to use them to their advantage for getting attention from humans. Every human language has a representation of this type of call: 'the English cat "mews", the Indian cat "myaus", the Chinese cat says "mio", the Arabian cat "naoua" and the Egyptian cat "mau". To illustrate how difficult it is to interpret the cat's language, her 'mew' is spelled in 31 different ways, five examples being maeow, me-ow, mieaou, mouw and murr-raow' (Moelk, 1944).

All, including the trill or greeting-miaow, which is often performed without opening the mouth, are uttered in amicable social encounters, to establish contact and incite further friendly interaction such as play or sharing food. The 'chirrup' form of the trill is particularly used by the mother as a contact call to her kittens, which by about 21 days of age are capable of distinguishing between their own mother's call and that of another female; both mother and kittens may also have an ultrasonic call that fulfils the same general function. The long miaow is a high-intensity version of the ordinary miaow, well known to owners of cats when they do not produce food quickly enough, but its function in cat–cat interactions is obscure.

Domestic cats produce shorter, higher-pitched miaows than their wild ancestors (*F. lybica*) and people, even those with no experience of cats, describe domestic cat miaows as far more pleasant sounding than wildcat vocalizations (Nicastro, 2004). Yeon *et al.* (2011) found that house cats also had different (shorter, higher-pitched) miaows from feral cats. An element of plasticity in this behaviour is suggested by these studies, and also by other studies using operant conditioning indicating that cats can learn to alter the fundamental frequency of their miaows, their duration and their call rate (see Nicastro, 2004). In addition miaows will often include clicks, trills, growls and other elements that may enable a listener to distinguish different calls.

That said, research suggests that despite the potential advantage to cats of conveying specific messages in their miaows, humans find it difficult to derive specific information from calls of unfamiliar cats. When asked to classify miaows from 12 unfamiliar cats in five different call-production contexts (food-related, agonistic, affiliative, obstacle or distress) humans could differentiate them, but not especially accurately. Classification of calls was more successful for those people that had lived with, interacted with and had a general affinity for cats, although only with respect to agonistic and affiliative calls. It is possible that with repeated interaction over time cats can learn to incorporate special context-specific acoustic cues into their calls that will be recognized by their individual owners as having a particular meaning. Known as 'ontogenetic ritualization', this theory is still being investigated in cats (see Nicastro and Owren, 2003).

Finally, some cats also have a soundless version of this call (i.e. a visual display), in which the mouth is opened and closed at the same rate as a miaow; often referred to as 'the silent miaow', this evidently has appeal to owners but its role, if any, in cat–cat behaviour is unknown, although it is also performed between farm cats.

Strained-intensity calling

Male and female feral cats produce loud 'rutting' cries specific to the breeding season, males producing them more often than females. Both sexes also produce distinguishable yowls and mews that are not specific to the breeding season (Shimizu, 2001). The yowl or 'caterwaul' can be emitted on its own or in combination with growling. It is probably no coincidence that the growl is one of the lowest-pitched calls used by the cat. Low-pitched sounds are generally characteristic of large animals, and so in aggressive encounters each cat may be trying to deceive the other into believing that it is larger than it really is. The hiss is primarily a defensive sound that proceeds to actual aggression only if the other cat presses home its attack. The other defensive threat, the spit, is the only call the cat uses that starts abruptly and at full intensity; it is probably the vocalization that is used most often to deter predators, and as such needs to be an unambiguous signal interpretable by species other than the cat.

Kitten calls and the development of adult calls

For the few days after its birth, the kitten has only two basic types of vocalization – a distress call, by which its mother is guided to attend to it, and the defensive spit. The distress call is given when the kitten is hungry, when it is trapped (for example, if the mother inadvertently lies on top of it) or if it becomes isolated and cold (Haskins, 1979). The rates of vocalization that these circumstances produce vary with the kitten's age, as the probability and consequences of each event change. Restraint always elicits a high rate of calling in kittens up to 6 weeks old, but kittens exposed to cold cry less as they get older and more able to regulate their own body temperature. Cold elicits a call of higher frequency and shorter duration than occurs in other circumstances, and it may be that the mother can distinguish two or more types of the kitten call, and tailor her actions to suit. For example, the 'cold' call may stimulate her to retrieve a young kitten that has clung for too long to her nipple, and has dropped off well outside the nest. The effect of isolation from the mother (without cooling) peaks in the third week, presumably once the kitten has formed a mental representation of the social group in which it finds itself. Movements of both the mother and the kitten's littermates can stimulate low levels of crying, but kittens that are huddled together and not disturbing one another hardly ever cry. Even in a strange environment, the presence of littermates reduces the level of crying; this is presumably an adaptation to prevent the kittens giving their position away immediately after their mother has moved them to a new nest site.

The response of the mother to the normal kitten call is context-dependent. If the sound comes from a location away from the nest, she is stimulated to leave the nest and investigate. If the sound comes from inside the nest, and she is herself outside, she will return to the nest and enter it. If she is inside, and with the kittens, she will usually sit down, and then lie down, allowing them to suckle. If for some reason one or more kittens are unable to locate a nipple, they may continue to cry, in which case the mother is further stimulated to shift position, jerking her abdomen so that the nipples are better exposed (Haskins, 1977). As each kitten grows, its call takes on characteristics that are individual-specific, but it is not known whether the mother can recognize each member of her litter by their cries.

From the time they are a few days old kittens can also purr, and do so particularly when they are suckling. In this situation, the sound will be transmitted to the mother and the littermates as much by direct skin contact as through the air. The purr may signify to the mother that the kitten is receiving a satisfactory amount of food. As the kitten grows, it may associate situations of amicable contact with its own, its littermates' and its mother's purring, and carry this association through to amicable contacts in adulthood, thus explaining the retention of purring throughout life.

Up to about 3 to 4 weeks of age, a kitten in pain will emit a high-intensity version of the kitten call, as well as spitting, but from about that time onwards the adult form of the pain shriek begins to emerge in its characteristic form. The kitten call itself is common until the kitten is 2 or 3 months old, but then declines until it is rarely heard from kittens over 5 months old. Miaows first appear soon after weaning, and at that stage are more likely to be directed at a human providing food than towards other cats. At about the same time, the kittens begin to snarl at one another, particularly in contests over food, and gradually the growl and anger wail are added to the repertoire of aggressive calls. The defensive hiss is absent until the kittens are at least 1 month old (Brown *et al.*, 1978).

Some of the changes that take place in the characteristics of the kitten's calls, such as their frequency, can be ascribed simply to the physical development of the vocal apparatus. Others, such as the duration of the call, tend to decline up to the end of week 3, probably reflecting the kitten's growing independence at this stage, and therefore a reduced need to call for its mother's attention. Because kittens are virtually deaf when born, it is not surprising that the initial form of the kitten call does not depend on auditory experience in any way. However, from about day 10 onwards, normal kittens control the loudness and frequency range of their calls by listening, both to themselves and to littermates. By 3 weeks of age, the calls of deaf kittens tend to be distinctly louder and lower pitched than those of normal kittens (Romand and Ehret, 1984).

Visual Signals

The evolutionary history of the cat as a solitary predator means that it lacks the complex visual signalling repertoire typical of social predators. Rapid changes in facial and body postures, and the ability to interpret these signals equally rapidly, are important in social species to coordinate shared activities and avoid conflict over shared resources. Hence Carnivora that have evolved as social predators, such as the wolf, have a sophisticated ability to display changes in emotional state, and to identify such changes in others in order to appropriately modify their behaviour (Bradshaw *et al.*, 2009). Since the evolution of the cat prior to domestication did not select for such abilities, visual signalling is limited in both scope and complexity. It is also likely that the ability to rapidly recognize and learn about the signals of others (i.e. social cognition) is less well advanced in this species than in more social carnivores. Hence, while cats are able to show a range of visual signals, these are less subtle in nature and tend to be mainly associated with defensive reactions rather than social integration or affiliation. Investigating the function of visual signals is much more difficult than are sounds, which can be recorded and played back, or scent marks, which can be moved from place to place or even extracted and re-deposited on a new surface. Their effect can usually only be judged from the reactions of other cats during social interactions, and as such their functions can only be inferred. Paul Leyhausen (1979) described and categorized in great detail the body postures associated with agonistic behaviours, and much of the account that follows stems from his work, and also from an ethogram for behavioural studies of the domestic cat compiled by the UK Cat Behaviour Working Group (1995). Many of these behaviour patterns are signals that form the basis of social interactions in groups of cats, which are discussed in detail in Chapter 8.

Whole-body signals

Most whole-body signals in cats are associated with defensive responses to per-ceived threats. The type of response varies between situations in which cats take an aggressive or avoidant strategy to threat, presumably based on the relative suc-cess of each in similar previous situations. Like many other mammals, cats change their apparent size during aggressive encounters. An attacking cat will cause its fur to stand on end (piloerection) and draw itself up to its full height; only the external ears are drawn back, presumably because these are often damaged in fights, and can subsequently become badly infected (Fig. 5.1, top right). A defensive cat tak-ing an avoiding strategy will crouch on the ground, flattening its ears and coat, and generally try to look as small, and presumably as unthreatening, as possible (Fig. 5.1, bottom left). Its head is drawn into its shoulders, indicating that it is not in a position to launch an accurate biting attack. Continued threat to such a cat may lead to a change in response to aggression, usually starting with a swipe of the paw. Appeasement signals, commonly seen in the dog, do not appear to occur in cats, making conflict resolution more difficult (see Chapter 8, this volume). In the dog, appeasement postures vary from subtle signs such as gaze aversion to overt signals such as rolling over to expose the inguinal region. Leyhausen's book (1979) contains drawings of a cat that has backed into a corner and is perform-ing a near-headstand, while still swiping at its aggressor with its front paws, but this is likely to be a defensive response rather an appeasement response to resolve conflict. Leyhausen (1979) considers a third agonistic posture, the arched back, to be the result of conflicting tendencies to attack and defend (Fig. 5.1, bot-tom right), and has illustrated a number of intermediate postures that connect all three extreme agonistic positions (Fig. 5.1, middle rows). A cat in the arched back posture often stands sideways on to its opponent, which may be another cat, a

Fig. 5.1. The whole-body postures that signify anger (increasing from left to right) and fear (increasing from top to bottom). See text for further explanation (redrawn from Leyhausen, 1979).

dog or, if the cat is an unsocialized feral, a human. A similar posture is a common component of play in the kitten (see Chapter 4, this volume).

The use of body language in non-agonistic encounters and its role in maintaining social structure within groups has received attention in recent years and is described in more detail in Chapter 8. Females in oestrus go through a sequence of postures that includes rolling on to the back and on to the feet again. Some adult males also roll over on to their backs, apparently as a form of display, but rarely, if ever, as part of agonistic encounters. High-intensity displays of object-rubbing (described below) may be preceded or followed by rolling, suggesting that some sort of transfer of scent may be taking place, but in other instances rolling can form part of social encounters. This display may therefore have a meaning outside sexual behaviour, but what this might be is not clear. It also forms part of the catnip response (see below).

Facial expressions

In addition to its body posture, a fearful cat will often deliberately avoid eye contact with its aggressor, by pointing its head to the side. Anthropomorphically, this might be interpreted as the fearful cat 'taking no notice' of the other (i.e. actively avoiding the confrontation), but the escalation of violence that ensues if the cats do make eye contact shows that the individual that is looking away is actually highly aware of the other. Cats appear to minimize the risk of confrontation by deliberately not making eye contact. The ears provide the most obvious cues as to the cat's intentions – they are folded sideways and downwards as part of the avoidance response. This should not be confused with backwards rotation, which is part of the aggressive posture. Ear positions can be altered much more quickly than whole-body postures, and so the two do not necessarily change precisely in parallel – facial expressions can alter dramatically while the body maintains a fixed position. It would be surprising if ear and eye signals were not also used in the context of affiliative encounters, but these have not been investigated in detail.

Tail signals

The tail is fundamentally an organ of locomotion, and cats that are jumping, climbing or galloping use their tail for balance. At slower gaits or while standing the tail is available to provide signals, and is unusually expressive because the tip can be moved independently of the base. In fact, there has been very little study of the meaning of these tip-twitches, but some contexts that can give rise to the basic whole-tail positions have been documented (see Table 5.3). The normal tail position is horizontal or half-lowered; the most

Table 5.3. Tail positions used by the adult cat, and some circumstances under which each typically appears, indicating possible roles in communication (from Kiley-Worthington, 1976; Brown, 1993).

Tail position	Circumstances
Vertical	Greeting (walking, trotting or standing)
	Social play; object play
	Sexual approach (female)
Half-raised	Sexual approach (female)
Horizontal	Amicable approach
	Sexual approach (female)
Concave	Defensive aggression
Lower	Aggression
Between legs	Extreme fear

obvious signals are produced when it is raised to the vertical, as a 'greeting signal', or raised at the base and then curved progressively downwards to form the concave position (Fig. 5.1, posture immediately above the bottom right-hand corner).

The displaying of a vertically held tail or 'tail up' in cats is associated with affiliative behaviour (Brown, 1993; Bernstein and Strack, 1996; Bradshaw and Cameron-Beaumont, 2000). Kittens display the tail-up when greeting their mother, and this behaviour has also been observed in the wild ancestor. In the domestic cat, tail-up can also be seen when one adult individual meets another. Cameron-Beaumont (see Bradshaw and Cameron-Beaumont, 2000) found that cats presented with the silhouette of a cat with its tail up were more likely to raise their own tail and also approached faster than if the silhouette had its tail horizontal. In contrast, tail-down silhouettes induced aggressive responses in some cases, suggesting that tail-up has the function of inhibiting intraspecific aggressive behaviour. The role of this behaviour in cat social interactions is discussed further in Chapter 8.

The defensive threat posture (Fig. 5.1, bottom right) was drawn by Leyhausen (1979) with the tail erect, but other authors have indicated that a concave tail is more typically associated with the arched back. During the preliminaries to actual fights, the tail is kept out of the way, presumably because it is susceptible to damage. The aggressive cat points its tail at the ground, and the tail is lowered in cats showing avoidant responses.

The tail can also be moved from side to side, and one obvious signal produced in this way is the aggressive tail-lash that often accompanies growling. Less exaggerated movements can be observed under a wide variety of circumstances, such as in situations appearing to involve frustration or indecision, but their importance in communication is unknown.

Olfactory Communication

Most carnivores lead a solitary existence for the majority of their lives, and therefore often find it difficult to predict whether visual or acoustic signals will be detected by any conspecific. Odorous secretions have the advantage that they can be left to provide information about their producer for many days after they were deposited. Because for many years domestic cats were thought to be asocial, most studies of their olfactory communication have been interpreted as if all cats were solitary. It is likely that odours play all kinds of roles within social groups, but not all of these are well understood. For example, it is not certain whether colonies of cats share a common 'colony odour' identifying them as belonging to a particular group, in addition to odours that may be individual-specific, or could give information on sex, age or status. Other social carnivores do have such systems – for example, badgers exchange the contents of a large, pocket-shaped gland beneath their tails, the subcaudal gland, to produce a 'clan odour' (Buesching *et al.*, 2003).

Being warm-blooded, mammals are inevitably presented with the problem that their scent materials will become contaminated by microorganisms. The scents are often produced from modified skin glands, and consist of lipid- and protein-rich materials on which a variety of bacteria and yeasts can grow, producing their own odours. In some cases, the mammal appears to regulate the microflora by producing the secretion into a specialized structure, such as the anal sacs of the domestic dog. In other cases, for example when the scent-producing material is voided on to hairy skin, there can be little control, and there will be even less once the scent mark has been deposited. How an effective communication system can work under such a set of uncertainties is still a mystery, but it has been suggested that so long as the microflora consists of a reasonably stable group of species, each animal should be able to produce an individual-specific odour that can be learned by others. The production of group-specific odours must involve not only the mixing of scented material, but also of microorganisms, which is presumably why badgers place such emphasis on exchanging the contents of their subcaudal pouches.

In addition to skin glands, the vast majority of mammals use their urine and faeces for odour communication, either relying on their intrinsic odours or using them as carriers for glandular secretions. Urine marking has been studied extensively in domestic cats, faecal marking less so. The use of skin glands has attracted more studies in recent years as scientists have been able to artificially reproduce some of the secretions for use in behavioural therapy.

Urine

Tomcats are renowned for their urine spraying, an unmistakeable and, in a pet, rather undesirable piece of behaviour. The cat backs up to an object and stands, tail erect and quivering, while urine is sprayed out backwards and upwards on to the object (Fig. 5.2). Females will also spray urine, though not as frequently as toms, and neutering does not always prevent the development or even the onset of spraying. In closed or high-density colonies there may be some suppression of spraying in females and younger males, resulting in the majority of spray marks being produced by one or a few older, less inhibited, males. For example, in one group of six females and four males, only one male regularly sprayed (Natoli, 1985). Both males and females tend to increase their rate of spraying when the females are in oestrus, producing a seasonal pattern that peaks in February/March.

In a mixed-sex farm colony, males sprayed most often when in consort with an oestrous female, less often but still frequently (about once every 10 min) when hunting and least often when in the farmyard, where they would usually be in the company of other cats. While out hunting, they directed their sprays at visually conspicuous sites, such as the sides of buildings, hay bales, fallen branches, fenceposts, grass tussocks and molehills. Spraying was not seen immediately before, or during, bouts of male–male aggression, although other

Fig. 5.2. Urine spraying.

carnivores – domestic dogs, for example – often deposit urine during competitive encounters. Spraying by males can form part of courtship behaviour (see Chapter 8, this volume).

Individual differences in the rate of urine spraying have been noted. For example, Feldman (1994b) observed between 2.8 and 9.2 sprays per hour by the males in her study, possibly reflecting differences in either age, territoriality or individual learnt strategies to social conflict. Spray marks do not appear to act as a deterrent in their own right, since cats have rarely been seen to leave or avoid an area after investigating one. They possibly serve more as information markers (Bateson and Turner, 1988; Feldman, 1994b), for example to enable cats to avoid contact and potential conflict. Feldman also found that spray marks were not concentrated at the periphery of the ranges as would be expected if territory marking were the primary function of spraying. So the territorial function of urine spraying, if any, remains unclear (Bradshaw and Cameron Beaumont, 2000).

Young cats of both sexes, adult females, neuters and, to some extent, adult males also urinate in the squatting position. Squat urinations are usually buried as if the cat were trying to hide whatever information might be conveyed by the smell. It is thought that such urinations are primarily eliminatory in function, but squat urinations, if detected, are sniffed by both male and female cats. However, both sexes sniff sprayed urine for longer than squat urine, so it appears likely that the two types contain different information, and that sprayed urine possibly conveys more detail. The source of this extra information has not been investigated in detail, but there do seem to be chemical differences between squat and spray urine, in addition to the deliberate targeting of the latter. Some authors (e.g. Wolski, 1982) have claimed that spray urine is cloudy, and contains droplets of a viscous lipid material that might originate in the anal sacs. By comparison with most other carnivores, these are small and unspecialized, and their role in communication has scarcely been investigated. Their secretion is usually yellow–brown in colour and has a characteristic odour, and it is likely, since this is known to occur in male lions and tigers, that it could be discharged into the stream of sprayed urine. On the other hand, tomcats can produce a clear, strong-smelling liquid from beneath their tails, and this may add to the strong odour of the sprayed urine of intact males.

Among pet cats, spray-marking is most common in males living in multi-cat households. According to owner accounts, these cats most often mark as part of agonistic interactions with other cats, either those in the home or intruders from other households (Pryor *et al.*, 2001). However, owners are presumably much more likely to notice spraying when it occurs inside the home than outside, so further research is needed to establish the reasons for and locations of spray marking by pet cats.

Information conveyed by urine marking

The information conveyed by cats' urine marks has not been investigated in detail. One rather crude measure that has been used widely in behavioural studies of urine marking is the amount of time cats spend investigating a given sample. Males generally spend rather longer at this than females, unless the female is in oestrus, when she will pay great attention to the sprayed urine of males, particularly if they are strangers. Correspondingly, males sniff oestrous female urine for longer than anoestrous. Males also appear to discriminate better between urine samples from unknown cats, cats from neighbouring groups and cats of their own group (Fig. 5.3), although females also indicate that they can make this distinction. This does not necessarily mean that each cat can recognize each urine scent mark as belonging to a particular individual, although this is possible, as has been shown for the European lynx (Sokolov *et al.*, 1996). In social groups of cats there is also the possibility of 'clan odours', either actively maintained by transfer of odour materials or odour-producing microflora between individuals (as described above for badgers) or passively, for example by the sharing of food sources with characteristic odours within, but not between, groups. Urine is not only sniffed, it can also induce the Flehmen or 'gape' reaction in which

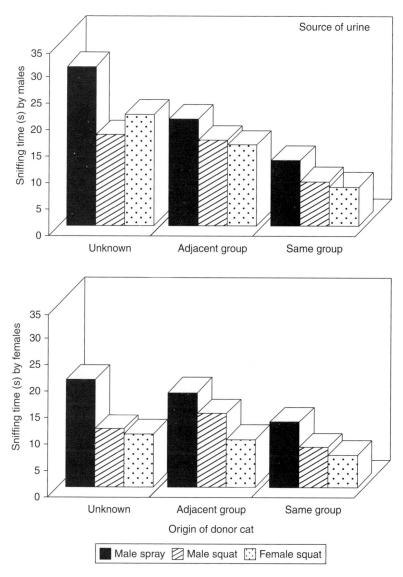

Fig. 5.3. The amount of investigation induced by three types of urine from three classes of cat, in male (top) and female (bottom) farm cats (data from Passanisi and Macdonald, 1990).

the vomeronasal organ is activated (see Chapter 2, this volume), although the extent to which this additional form of olfactory inspection is used seems to vary considerably from cat to cat and is possibly commonest towards the scent mark of a strange male. While all cats will initially sniff a urine spray mark, many follow this with bouts of Flehmen, in which the tongue is moved rhythmically along the roof of the mouth, and then the head is raised and the mouth opened, a posture which is maintained for up to a quarter of a minute. Head shaking may occur between bouts of Flehmen. Sniffing at a distance, when the cat behaves as if it is then attracted by the odour of the urine, is only common when the urine sample is fresh; older marks are often detected because cats tend to inspect the prominent objects on which they were deposited. Prolonged sniffing of old scent marks can result in the mark being moistened by the cat's nose, and Flehmen will sometimes follow, although less commonly than to fresh marks.

Before it became clear that scent marks of all kinds might play a role in regulating social behaviour, it was thought that the primary function of spray-urination was to enable hunting cats to space themselves apart. The hypothesis was that by avoiding areas that had been recently hunted by others, each cat could maximize its chances of encountering undisturbed prey and minimize the probability of an encounter with another, possibly hostile, cat. The odour of a scent mark changes as it ages, due to a combination of differential evaporation of volatiles and the production of new smells by microorganisms, and could be used to indicate the time since a cat last hunted that area. Certainly, foraging cats do spend a great deal of time investigating scent marks, but their spray-urinations tend to be concentrated in their core area; at the edges of their home ranges their rate of urination tends to drop, which would not be predicted by the time-sharing hypothesis. Individual cats may indeed make use of the fresh urine marks to inform themselves of the likely whereabouts of others, but it is likely that these marks also confer some advantage on their emitter.

The potential information conveyed by a urine mark, based on its chemical composition, has received some attention in recent years. Cat urine contains unusual sulfur-containing amino acids, most notably felinine, which is gener-ated by cauxin, an enzyme produced in the kidneys that is unique to small cats (Miyazaki *et al.*, 2008). These have little odour in their own right, but their breakdown products (including 3-mercapto-3-methyl-1-butanol) are highly pun-gent and contribute to the odour of urine marks as they age. Microbes carried along with the urine as it is voided are likely to increase both the strength and character of the odour.

The composition of feline urine varies according to gender, age, health and reproductive status (see Miyazaki *et al.*, 2008). Cauxin, regulated by the level of testosterone in the blood, increases the urinary concentration of felinine. The highest urinary concentration of felinine and cauxin is found in mature intact male domestic cats – females and castrated and immature males excrete substantially less. Felinine is biosynthesized from cysteine and methionine, both important amino acids in the cat's diet (see Chapter 6, this volume), and mature tomcats excrete such quantities that the dietary requirement for these amino acids is significantly increased (Hendriks *et al.*, 1995). Since muscle meat is the cat's main dietary source of cysteine and methionine, and the odour of tom-cat urine is largely characterized by sulfur-containing breakdown products of felinine, the strength of that odour may be an 'honest signal' of prey-catching ability, hypothetically available to oestrous females for assessment of competing males, or to males attempting to assess the likely competitive abilities of other males. The use of spray marking during agonistic encounters between males with adjacent territories (Pryor *et al.*, 2001) is consistent with the latter hypothesis.

Faeces

Many carnivores use their faeces, sometimes with glandular secretions added, to disseminate olfactory information. For example, European otters deposit piles of faeces, known as 'sprants', at nose height, throughout their home ranges on the tops of prominent objects such as large rocks, where they can be easily detected by other otters. The role, if any, of faecal odours in cat–cat communication is still not certain. Farm cats usually have communal latrine sites, which tend to be in areas with a loose substratum, such as turned soil, gravel, sand or hay. Burying of faeces is common at these sites, although some stools are always exposed, either because no attempt has been made to bury them or because in burying one defecation another has been uncovered. The significance of such communal sites may rest on the scarcity of areas with suitable substratum, rather than being connected in any way to communication. Away from the farmyard, piles of faeces can occasionally be found on exposed, conspicuous sites. If the primary function of covering is one of hygiene, then it may not be necessary away from the centre of a communal territory, but the use of obvious sites cannot be explained in this way and suggests some sort of communicative function.

Ishida and Shimizu (1998) found that although heavier male cats did not expose their faeces more frequently than lighter cats, when they did bury them it tended to be at sites closer to the core area than lighter cats. No such relationship was found for females. It is possible that perception of social threat to their core area may therefore in some way influence defecation behaviour in male cats. In general, however, patterns of burying or leaving faeces exposed have proved variable between studies of different cat populations and the role of faeces as a marking behaviour remains inconclusive.

Scratching

Within a cat's home range, certain objects are marked repeatedly by being scratched with the claws of the front feet (Fig. 5.4). Favoured sites are usually vertical wooden surfaces, such as the trunk of a tree or the side of a shed, and these can become deeply grooved if used over a long period, providing a potential visual signal. Fragments of claw and claw sheath can often be found embedded in these grooves, giving the impression that the primary purpose of the scratching is to condition the claws. However, if this were the case, then presumably there would be at least some need to scratch with the claws of

Fig. 5.4. Scratch marking.

the hind feet as with those of the front. The scratched sites presumably act as combined visual and scent signals, the latter derived from the sebaceous glands of the feet. Although detailed information on these secretions is lacking, Pageat and Gaultier (2003) note that sweat secreted by fearful cats enhances avoidance behaviours in cats encountering it.

In a study of scent marking by domestic cats, Feldman (1994b) observed that trees along paths were scratched more than trees along the perimeter of their range, suggesting that domestic cats – like other carnivore species – mark within their range along habitual paths rather than defining the edges of their territory with scratch marks.

Tree characteristics also appear to influence the distribution of scratches, with soft-barked trees being scratched more than hard-barked trees. Harder bark, providing more resistance, would presumably be less suitable for visual marking (Feldman, 1994b).

The precise role of these signals, and their relative importance compared with other scent marks, such as urine and faeces, is unknown. Feral cats perform claw sharpening more often in the presence of conspecifics than when alone, suggesting that the action itself may serve as a visual gesture of dominance (Turner, 1988).

Skin glands

Cats have a variety of specialized skin glands, several of which appear to be communicatory in function. There is a large submandibular gland beneath the chin, in the area where most other carnivores have a tuft of whiskers, but it produces no obvious secretion and most cat owners probably never know of its existence unless it becomes infected. There are perioral glands at the corners of the mouth, temporal glands on each side of the forehead and diffuse clumps of sebaceous glands along the tail, collectively known as the caudal glands (Wolski, 1982). Other skin glands, at the base of the tail, enlarge as the cat matures, particularly in intact males, and may be the source of the clear liquid mentioned above. The cheeks and ears may also produce odorous secretions, and the mutual flank-rubbing display performed by familiar cats suggests that this area also may be endowed with scent-producing structures.

The secretions of the submandibular, cheek and perioral glands can be deposited as scent marks on prominent objects at head height, such as twigs that project over frequently used paths (Fig. 5.5). Also known as bunting, this

Fig. 5.5. Scent marking by cheek rubbing.

marking behaviour can be performed in isolation or following Flehmen towards a urine mark or head mark, and may itself be followed by spraying if the marker is a tom. The precise form that head rubbing takes seems to depend at least partly on the topography of the object being rubbed. Cheek rubbing is performed along a line from the corner of the mouth to the ear. Higher objects may be marked on their undersurface using the forehead and ears; objects near the ground may be marked first with the underside of the chin, then with the side of the throat (Verberne and De Boer, 1976). Inanimate objects may also be rubbed with the flanks and the tail, although these areas of the body are more commonly used in cat–cat rubbing.

Solitary cats pay a great deal of attention to the sites of previous head marks when they are patrolling their home ranges. In one group of farm cats, Warner Passanisi found that females increased their rate of head marking during oestrus, while males showed a similar increase when consorting with a female (Passanisi and MacDonald, 1990). Males would occasionally head mark away from their core area, but females never did. Females may spend longer than males in investigating head marks in general, but males pay particular attention to head marks left by oestrous females. Male domestic cats can differentiate phases of the oestrous cycle of females from cheek gland secretions (Verberne and DeBoer, 1976).

Mellen (1993) suggests cheek rubbing in small felids serves three functions: to deposit a scent, to pick up scent and as a visual display (males do this when attending females in oestrus).

Five different facial pheromones (named F1 to F5) have now been isolated from the sebaceous secretions of the cheeks of domestic cats, with the functions of three of them elucidated to date (Table 5.4; Pageat and

Table 5.4. Chemical components of facial secretions in the cat (from Pageat and Gaultier, 2003).

Secretion	Components	Function(s)
F1	Oleic acid, caproic acid, trimethylamine 5-aminovaleric acid, n-butyric acid, α-methylbutyric acid	Unknown
F2	Oleic acid, palmitic acid, propionic acid, p-hydroxyphenylacetic acid	Sexual facial marking in males
F3	Oleic acid, azelaic acid, pimelic acid, palmitic acid	Facial marking on items (synthetic version available)
F4	5β-Cholestan acid 3β-ol, oleic acid, pimelic acid, n-butyric acid	Allomarking (synthetic version available)
F5	Palmitic acid, isobutyric acid, 5-aminovaleric acid, n-butyric acid, α-methylbutyric acid, trimethylamine, azelaic acid, p-hydroxyphenylacetic acid	Unknown

Gaultier, 2003). F3 is deposited during object rubbing while a cat patrols its home range as described above. Male cats rubbing on prominent objects whilst performing a visual display towards a female in oestrus deposit the F2 pheromone. The F4 pheromone, known as the allomarking pheromone, is deposited when cats rub against each other socially (see below), and has been suggested to decrease aggressive behaviour between them, although further evidence is required to confirm such a role. The F3 and F4 pheromones have now been artificially synthesized and are used as aids in behavioural therapy (see Chapter 11, this volume).

Allorubbing

The mutual head, flank and tail rubbing of cats is discussed as an important component of social behaviour in Chapter 8. As mentioned above, scent is inevitably deposited when one individual head rubs another. When cats sniff one another they tend to concentrate on the facial and perineal areas, suggesting that any materials deposited on the flanks and tail by allorubbing may be less important as social cues than individual odours produced by the head and anal glands, and around the genitals. There may also be an element of picking up or sharing odours in addition to depositing them, and there is most likely some visual and tactile significance to this display. Within cat societies that contain breeding males, females and kittens, the initiation of rubbing is highly asymmetric.

Catnip

The response of some cats to the herb catnip (*Nepeta cataria*) – and to other plants containing compounds similar to nepetalactone, the most active compound from catnip – has been exploited for many years by the makers of cat toys, but the behavioural significance of the response is still not entirely clear. It can be assumed that the aroma of catnip conveys some message to cats, but no adaptive significance for this has ever been elucidated, and it must be assumed that the occurrence of the triggering compounds in a few types of plant is an evolutionary accident. The initial reaction usually involves components of object play, but in many cats this is followed by an intense combination of face rubbing and body rolling (Fig. 5.6) that can be elicited in males and females, both neutered and intact. Thus, while the rolling pattern is very similar to the same behaviour when it forms a part of female oestrous behaviour, the whole catnip response does not appear to be directly related to any sexual response, nor is it motivationally connected to hunting or aggressive behaviour (Palen and Goddard, 1966). The response is inherited as a dominant autosomal gene, and therefore not all cats show it (although the catnip herb is very variable in both the quality and quantity of its essential oil, possibly accounting for other variations in response).

Fig. 5.6. The catnip response.

Feeding Behaviour

6

Introduction

The function of feeding behaviour is to obtain adequate nutrition, and the nutritional requirements of the cat have not been changed during domestication – the machinery that supplies the body with energy and the building blocks for growth and repair is essentially that of a wild felid. Moreover, the whole cat family is nutritionally peculiar in a number of respects, most of which can be directly related to their carnivorous lifestyle. Of all the Carnivora, the felids are among the most specialized meat-eaters, known as 'hypercarnivores'. Despite the name of the group many Carnivora, including the bears, are actually omnivores and some – for example, the pandas – have secondarily reverted to plant-eating. This account of feeding behaviour will therefore start with a description of its goal, the fulfilment of nutritional requirements.

Nutrition

A meat diet is rich in protein and fat, but low in carbohydrate compared with a vegetarian diet; it has been estimated that the typical energy intake of a feral cat comprises 52% protein, 46% fat and only 2% carbohydrate (Plantinga *et al.*, 2011). Omnivores have a wide range of biochemical mechanisms for converting nutrients from both plant and animal sources into forms that they can use directly; obligate carnivores such as the cat are faced with a significantly narrower range of options. To take a simple example, the visual pigment in the retina, rhodopsin, is partly formed from retinol, also known as vitamin A. A similar group of compounds, the carotenes, are commonly found as yellow and orange pigments in fruits and vegetables, and most omnivores use these as precursors of vitamin A. Cats do not have this ability, and therefore need to obtain all their vitamin A as vitamin A itself, from animal sources. Strict specialization as meat-eaters appears to have resulted in the loss of certain metabolic functions, which now means that cats are unable to obtain a balanced diet from plant materials alone. Further examples of these restrictions will be discussed as each group of nutrients is considered. Cat nutrition is described in more detail by Morris (2002), Zoran (2002) and Zoran and Buffington (2011).

Protein

Cats require a much higher proportion of protein in their diet than almost any other mammal; the minimum level for body maintenance in the adult is about 12% (of dry weight of food) and growing kittens need about 18%. The main reason for this unusually high requirement is that the cat's liver contains highly active N-catabolic (protein-degrading) enzymes that cannot be shut off. Cats automatically take much of the glucose they need by gluconeogenesis from the protein part of their diet (Eisert, 2011), whereas the majority of mammals have the ability to direct almost all their dietary protein into growth and body main-tenance. For example, the protein requirements of the domestic dog (4% for an adult, 12% for a puppy) are much more typical of an omnivore. Note that the puppy requires three times as much as the adult dog, but the kitten requires only one-and-a-half times as much as the adult cat, reflecting the cat's under-lying need for protein as an energy source at all stages of life. Presumably, since the natural foods of cats and other specialized carnivores are always protein-rich, they have not suffered from losing the ability to conserve protein when it is in short supply: hypercarnivores from other taxa (including fish, reptiles and birds) have independently evolved similar adaptations to their protein-rich diet, indicating that these changes are actually adaptive.

All of these figures only apply to protein that contains the right blend of amino acids. The list of amino acids that are essential in the diet does not vary greatly between one mammal and another; in the cat these include arginine, histidine, lysine, leucine, isoleucine, valine, methionine, threonine, tryptophan and asparagine. The cat is, however, much more susceptible to a deficiency of arginine than other mammals. A single arginine-deficient meal can lead to ammonia intoxication, which includes emesis and lethargy among its immedi-ate symptoms. The missing arginine is required as an intermediate in the urea cycle, which is essential for the excretion of nitrogen, and when it is absent that nitrogen accumulates as ammonia in the blood. The reason that the cat is so susceptible is that, following each meal, cats deplete their levels of ornithine, another urea cycle intermediate that can substitute for arginine if the latter is in short supply. The depletion of ornithine prevents the sudden surge of amino acids in the bloodstream that follows the next meal from being broken down immediately to provide energy. Cats are virtually unable to make ornithine from any other source, and so require arginine at every meal to restart the urea cycle.

Cats also need unusually large amounts of sulfur-containing amino acids, such as cysteine and methionine, and it is these that are often the most limiting in normal foods. Because these amino acids are used in large quantities for making hair, the cat's thick coat has been suggested as one reason for this heavy requirement. Cysteine is also used for the synthesis of felinine, the unusual branched-chain amino acid found in cats' urine (see Chapter 5, this volume). Cats also need a β-amino acid, taurine (2-aminoethane sulfonic acid), which is not found in proteins, but is normally biosynthesized from cysteine. The enzyme responsible for this biosynthesis has very low activity in cats, resulting in a dietary requirement. Long-term taurine deficiency can bring about a degenerative disease of the heart muscles (dilated cardiomyopathy) and retinal degradation, and can also lead to poor reproductive performance.

Fat

In general, fat has three functions in the diet: providing energy, acting as a solvent for some vitamins and supplying essential fatty acids. Unsurprisingly for a carnivore, cats can make very efficient use of very high levels of fat in their diet, and moreover can digest fats that come from a wide variety of sources, including plant oils. One common adaptation to a carnivorous way of life appears to be low activity in one or more enzymes that produce arachidonic acid, itself a precursor of prostaglandins and other compounds essential to reproduction; such changes are found in animals as diverse as the lion, the turbot and the mosquito. Omnivores are capable of converting linoleic acid, a common constituent of plant oils, into arachidonic acid, and do not therefore require the latter in the diet. Without arachidonic acid in the diet, queen cats may not come into oestrus or, if they do, their kittens may show deficiencies in central nervous system development. The only known precursor of plant origin that cats can use is found in evening primrose oil.

Carbohydrate

Provided their diet contains both protein and triglycerides, cats do not need any carbohydrate. Many commercial cat foods, particularly dry formulations, contain significant amounts of carbohydrate, and cats are perfectly capable of digesting starches and most sugars, including glucose, although their capacity to utilize sucrose is limited. Like most carnivores they are incapable of digesting cellulose. One adaptation to their carnivorous habits is a lack of amylase in their saliva; another is reduced glucokinase activity in the liver. In herbivores and omnivores (including the domestic dog), this enzyme copes with the sudden increase in glucose that follows an easily assimilated, carbohydrate-rich meal. No such surge follows meat meals, so this enzyme is redundant in specialized carnivores. Cats also have inefficient lactase, which means that they are only able to digest low levels of lactose. Any excess passes to the hind-gut where it can be fermented by bacteria, sometimes causing diarrhoea. This accounts for some cats' intolerance of milk. Cats find milk fats highly palatable, but the presence of quantities of lactose in cow's milk means that for many this is not a suitable food.

Minerals and vitamins

The cat's mineral requirements are not much different from those of other mammals. In the wild, they must consume at least some of the bones of their prey, because meat alone is deficient in calcium and phosphorus. Some cats appear to be intolerant of high levels of magnesium in their diet; crystals of the magnesium salt struvite can form in the bladder giving rise to feline urological syndrome. One of the advantages of a carnivorous lifestyle is that all food is comparatively rich in sodium, essential for nervous and excretory functions. Most plants contain much less sodium than herbivores require, and so most mammals have a highly sensitive taste for salt and some will travel long distances to salt-licks to obtain sufficient of this important mineral. Both the cat and the dog are rather insensitive to the taste of salt, reflecting the likelihood that their diet will automatically contain enough sodium.

Cats have a few particular requirements for vitamins. Like the majority of mammals, they do not require vitamin C, but, as mentioned in the introduction to this section, they do have a requirement for vitamin A that cannot be met from plant-type precursors. Cats also make very little vitamin D, even when sun-bathing, as the cholesterol derivative from which vitamin D is made under the skin of other mammals, including humans, is destroyed by a hyperactive 7-dehydrocholesterol reductase. Niacin is also essential, because the cat can make very little of this vitamin from tryptophan (a competing biochemical pathway is very active). Thiamine is also required in abnormally large amounts, although, since this vitamin is mainly used in carbohydrate conversions, less is needed when the primary sources of energy are fat and protein. The raw flesh of fish contains thiaminases, enzymes which destroy thiamine, so a diet of uncooked fish can result in thiamine deficiency.

Digestion and water relations

The gut of the cat follows the typical carnivore pattern – the fore-gut is emphasized, the hind-gut is reduced and the colon and rectum are not obviously differentiated from one another. The caecum is small. Overall the gut length is only four times the body length, reflecting the high digestibility of much of the cat's food (Edney, 1988). Meat has high water and protein content compared with many other foods and, unless heat-stressed, cats can live on the moisture from meat (or commercial canned cat foods) alone, without drinking. The kidneys are capable of producing much higher concentrations of urea than those of man (2000 mM compared with <800 mM), which has been presumed to be an adaptation to the arid habitat of the ancestral populations of *Felis lybica*. For whatever reason, some cats drink only occasionally, although they must have water available if they are fed on dry or semi-moist cat foods. Whole milk, although a popular addition to many companion cats' diets, is less suitable as a source of water, partly because it contains substantial amounts of protein, fat and carbohydrate and is therefore a food in its own right, and partly because some cats are intolerant to lactose.

Taste

Cats undoubtedly use both the sense of taste and the sense of smell in selecting food, and physical characteristics such as texture and temperature can also be important. Olfaction has been discussed in Chapter 5, but its precise role in food selection has received little study. Cats probably use smell as a secondary cue in detecting their prey; hearing and sight are probably more important, as will be discussed in Chapter 7. Many owners will have noted their cats sniffing at an unfamiliar food before eating, and such sniffing may be followed by rejection (van den Bos *et al.*, 2000), but the whole flavour of a food, made up of a combination of its smell and its taste, is probably more important than odour, taste or texture alone in sustaining actual eating. For example, exposure to odour alone will not usually overcome neophobia to a strange-smelling food, whereas actually eating that food will do so rapidly. Some experiments with dogs have indicated that meaty odours alone will not sustain interest in an otherwise bland food, and it is likely that cats are also not easily misled in this way. The sense of taste has been studied in some detail in several domesticated animals, including the cat. Human taste sensations can be easily measured by asking subjects to rate their responses to standardized tastants; such reporting is of course unavailable when the subjects are animals, and so most of our information on animal taste responses comes from neurophysiological investigations (Boudreau, 1989). There are four cranial nerves that convey information on taste from the mouth to the brain, but of these only the facial nerve has been studied in detail. A few accounts describe the basic properties of taste-sensitive neurons in the other three (the glossopharyngeal, the vagus and the trigeminal), but these have tended to focus on their responses to just a few simple compounds. Several distinct taste systems have been identified in the facial nerve (Table 6.1), and the responsiveness of each has been established to a wide spectrum of taste compounds and nutrients, so that it is possible to make comparisons between the cat's abilities to taste particular types of compound and the nutrients that it is likely to encounter in its food.

Table 6.1. Summary of the major groups of taste neurons in the facial nerve (geniculate ganglion) of the cat, with their equivalents in the domestic dog, the rat and man included for comparison (adapted from Boudreau, 1989). M, major group; m, minor group; (m), imperfectly defined.

Neural group	Cat	Dog	Rat	Psychophysical equivalent (man)
Amino acid (cat type)	M	M	(m)[a]	Sweet/bitter
Acid (cat type)	M	M	–	Sour
Acid (rat type)	–	–	M	–
Nucleotide	M	M	–	Umami
Salt	–	–	M	Salty

[a]The amino acid groups in the rat are qualitatively different from those of the cat and the dog, and are predominantly located in the glossopharyngeal nerve.

Amino acid units

The taste buds are found both on fungiform papillae on the front, edges and top surface of the tongue, and also in four to six large, vallate (cup-shaped) papillae at the back of the tongue. Each taste bud contains 50–150 taste units that have neuron-like properties as well as being receptors (Reed *et al.*, 2006). The commonest type of taste unit in the facial nerve responds primarily to amino acids, and most of these originate in taste buds at the tip of the tongue. These units, in common with most of the others that will be described, produce a low rate of discharge spontaneously, and it is therefore possible for particular amino acids to cause an increase and/or decrease in the rate of discharge. By analogy with human taste descriptions, it seems likely that increases can produce one taste sensation, while decreases produce another, quite distinct sensation. Some amino acids trigger rapid rates of discharge in the amino acid units, while others (mainly those with hydrophobic side-chains) inhibit the discharge. The former category are generally described as 'sweet' by human tasters, and the latter as 'bitter', but of course this does not necessarily mean that cats experience the same subjective sensations that we do. It is known, however, that cats spontaneously prefer solutions of the 'sweet' amino acids, particularly proline, histidine and lysine, to plain water, and reject those in the 'bitter' group (White and Boudreau, 1975), so the analogy with our sweet taste at least may be a reasonable one. It is interesting that as far as their response to amino acids is concerned, these receptors are much more like those found in man than those of the rat. For example, L-arginine, which inhibits the cat units and tastes bitter to us, is highly stimulatory to the equivalent receptors on the tongue of the rat.

The same taste units also give a rather weak response to both common salt and potassium chloride, but cats seem to be much less sensitive to both of these compounds than are the majority of mammals, and show little or no discrimination between diets containing different amounts of salt, even if they are sodium-depleted (Yu *et al.*, 1997). Rats, for example, have a large population of highly salt-sensitive units in their facial nerves, for which no analogous taste buds have been found in the cat. However, many mammals also have a second population of salt-sensitive cells in the glossopharyngeal nerve, and since this nerve has not been investigated in detail in the cat the possibility of a salt-specific taste cannot be ruled out.

The main function of these units is presumably to give an impression of the amino acid profile of each prey item, or even of individual tissues within the larger species caught. Cats' well-known dislike of carrion may be accounted for by the accumulation of monophosphate nucleotides in many tissues after death; these compounds inhibit the amino acid units, and are actively rejected in the same way that the inhibitory amino acids are. Dogs, which eat carrion readily, have slightly different amino acid taste units that are inhibited by only a very few compounds.

The labelling of these units as 'sweet' automatically suggests that they should respond to sugars, and their equivalents in the domestic dog do indeed discharge vigorously when stimulated with a whole range of mono- and disaccharides, and even some artificial sweeteners. In the cat, neither this nor any of the other taste systems identified so far will respond to sugars presented at any behaviourally meaningful concentration. Cats will, in fact, drink solutions of sucrose in water as if they consisted of water alone, and if thirsty will go on drinking, with potentially disastrous consequences for their water balance (Carpenter, 1956). At the molecular level, it is now known that in cats one of the two genes that code for the proteins that make up the sweet receptor (*Tas1r2*) is defective – a pseudogene (Li *et al.*, 2005). The other (*Tas1r3*) is still expressed in taste buds, but its role in forming taste receptors is not currently understood for the cat, although by analogy with humans, it probably combines with a protein coded by *Tas1r1* to form the 'umami' taste receptor (see below).

Acid units

Our 'sour' taste is mirrored in the acid units of the cat. These respond to a whole range of strong and weak acids, such as nucleotide triphosphates, histidine, histidine dipeptides and protonated imidazoles, as well as carboxylic and phosphoric acids. Some of the low-molecular weight acids are preferred by cats, but medium-chain fatty acids, such as octanoic acid, are strongly disliked by cats. Even when these acids are presented as esters, enzymes in the saliva generate sufficient free acids to cause rejection (MacDonald *et al.*, 1985). Coconut oil, which is bland-tasting to us, must have a strong taste as far as cats are concerned, because it contains compounds that include medium-chain fatty acid moieties. This might explain some cats' dislike of hand creams and detergents that contain coconut and other plant oils, but its broader biological significance is obscure.

Raw meat contains only low concentrations of free fatty acids and is generally rather acidic in any case (pH 5.5–7.0), which means that many acids will not be in the ionized form necessary for the triggering of the taste response. Most of the stimulation of these units may come from the histidine dipeptides present. A few amino acids, most notably the sulfur amino acids L-cysteine and L-taurine, also trigger the acid units. At first sight it might seem that the response to taurine is an adaptation to the cat's unusual requirement for this compound (see above), but there is no behavioural evidence that cats reject diets deficient in taurine either immediately or even after repeated exposure. Furthermore, the acid units of the dog, which can make all its own taurine from cysteine, are more sensitive to taurine than those of the cat. This response, though measurable, may not have any particular biological significance beyond adding another component to the overall perception of flavour.

Other units

The remaining taste groups in the facial nerve are less well understood, but are all characterized by long latencies between stimulation and response, and irregular bursts of spontaneous discharge, rather than single spikes. All respond to nucleotide di- and triphosphates (Table 6.1), and are therefore analogous to the 'umami' units in humans (Reed et al., 2006). Additionally, rather poorly defined subgroups also respond to quinine (hence labelled the 'bitter' subgroup), tannic, malic and phytic acids and alkaloids. Cats are certainly very sensitive to quinine, rejecting it in solution at a thousand times the dilution rejected by rabbits or hamsters. The behavioural significance of these has not been established, but they do seem to be associated with carnivorous habits, because although they are uncommon or absent in herbivores, analogous systems are present in puffer fish and blood-sucking arthropods.

Meal Patterning

Cats are opportunistic feeders, and will vary their patterns of activity to take account of the availability of food, whether hunted, scavenged or provided by a human owner. As the difficulty of obtaining food increases, meal size increases, in a way that reflects not simply the cost of obtaining the previous meal, but the average difficulty of obtaining food over the past day or so (the 'global cost structure'; Collier et al., 1997). If palatable food is available all the time in a form that does not 'stale' – for example, a commercial dry food – cats will take small meals (typically 12–20) throughout both day and night (Mugford, 1977). The meals tend to be slightly larger during daylight hours than at night. The way that meal size and the interval between meals is determined is not fully understood, but over the course of a few hours the majority of cats are able to regulate their calorie intake to match their requirements, usually by decreasing the content of each meal rather than by reducing the number of meals. This contrasts with the situation in the majority of domestic dogs, which will overeat if fed *ad libitum*. House cats are, of course, capable of adapting to meal patterns imposed on them by their owners, and many pets are provided with food only two or three times each day, although the food may not all be consumed immediately, increasing the number of actual meals to, say, five or six.

Meal patterning may be a contributory factor to the growing proportion of cats classified as obese (German, 2006), since *ad libitum* access to food, and taking many small meals, are two of the many risk factors that have been identified (Russell *et al.*, 2000). It is also possible that some owners misinterpret their cats' attempts to initiate social contact (see Chapter 9, this volume) as requests for food, and as a consequence provide excessive amounts of highly palatable food.

Learning about Food

Because cats are specialized carnivores, it is entirely feasible that they could rely more on inherited notions of what is edible and what is not than on the complex mechanisms of food learning used by omnivores. In fact, as will be described in more detail in the following chapter, feral cats eat a wide variety of prey that varies with the seasons and can include mice, voles, rabbits, rats, birds, lizards and insects as well as food scavenged from human sources. House cats eat a range of commercial products, as well as various processed and cooked foods primarily intended for human consumption. It seems unlikely that information on the nutritional content of each one of these could be genetically programmed, and in recent years it has been shown that cats can indeed learn a great deal about the consequences of eating particular foods (Bradshaw, 2006).

To put these abilities into context, it is worth making some comparisons between species with different feeding habits. Eating generally depends on two criteria being satisfied: that the potential food is indeed food and that, if eaten, it will satisfy some internal need state (Rozin, 1976). For some animals, filter-feeders for example, internal states seem relatively unimportant and feeding always takes place if recognizable food is available. At the opposite extreme, opportunistic omnivores may ingest any combination of a very large number of possible foods, each of which may offer a different blend of nutrients, some matching current needs and some not. Specialist carnivores feed on a few nutritionally interchangeable foods, one example being the lions of the African savannah, and might hypothetically require only one need state, that of general hunger, since a single meal will satisfy all their nutritional needs; only vitamin-deficient zebras can produce vitamin-deficient lions. However, recent research has indicated that some members of the Carnivora, including cats (Hewson-Hughes *et al.*, 2011), do indeed select food based on its macronutrient content.

Many species behave in a relatively inflexible way towards two key nutrients, water and sodium. Hunger for water, so distinctive that we call it thirst, arouses a search for fluids. Omnivores and herbivores also have an innate preference for salty tastes, activated when there is an internal need for sodium.

Hunger for calories, usually called just plain hunger, is also innate, but since calories can come at different levels and proportions from fat, protein or carbohydrate in different foods, experience plays a major role in modifying meal sizes to take account of their calorie content: an association between the flavour of a food and its energy content is learned, and used subsequently in regulating the intake of that food.

Intakes of all other nutrients, and components of foods that should be avoided, are based on learning mechanisms. For many years it was unclear how learned mechanisms could possibly play a part, because experimental psychologists believed that associations could only be learned between events that took place simultaneously or very nearly so. It is now clear that, whatever the restrictions on other types of learning (see Chapter 3, this volume), learning about foods can and almost invariably does take place between sensory stimuli and physiological consequences that occur up to several hours later. For this type of learning, it is important that the sensory stimuli are biologically relevant. Taste, and possibly odour, is readily learned, whereas visual or auditory stimuli are not. For example, a thiamine-deficient diet rapidly becomes aversive to rats; when it is presented on a subsequent occasion, the rats approach it eagerly, sniff it, but then do not eat it. Foods with the same smell as the deficient food are also avoided, even those that are not themselves deficient, and this avoidance persists after thiamine balance is restored. In fact, the rats behave towards vitamin-deficient diets just as if they were slow poisons. The general principle that lies behind this sort of learning seems to be 'avoid the most relevant stimuli associated with disadvantageous gastrointestinal consequences'. Within the mixed diet typical of an opportunist omnivore like the rat, the association is generally established between the most recently eaten new food and the unpleasant consequence that followed, on the assumption that foods that have proved themselves to be safe in the past are much less likely to have contributed to the malaise than one of which the rat has no prior experience.

Rats can therefore learn from their mistakes. The number of mistakes that they make is kept to a minimum by neophobia, a distrust of new foods in general, which can become stronger following several unpleasant experiences. However, if they are actually in a nutrient-deficient state, rats faced with a choice between a food that they now know is deficient, and an unfamiliar food, will often choose the latter.

Despite the relatively restricted range of foods accepted as such by cats, compared with the catholic tastes of rats, most of these mechanisms occur in both species, although the cat has been investigated far less thoroughly.

Learning about flavours

Cats are generally quite ready to accept novel foods. The extent to which young cats are distrustful of new foodstuffs may depend in part on the types of food that their mother introduced them to when they were kittens. The 'personality' of the kitten may also be important, because a tendency to neophobia may be linked to general nervousness. Kittens raised on commercially prepared foods generally show the opposite trend – that is, they prefer new types to the brand they have been raised on (Mugford, 1977). As they become adult, most cats' food preferences tend to become somewhat fixed, probably based upon what their owners feed them, and bear little resemblance to the preferences they displayed soon after weaning (Bradshaw, 2006).

It can sometimes prove difficult to introduce new foods into the diet of adult cats, and in some instances neophobia is the major cause of refusal, although it is entirely possible that the cat simply finds the food unpalatable. Perhaps surprisingly, many cats maintained on commercially prepared foods are neophobic towards raw muscle meat (Bradshaw *et al.*, 2000b). Depending on the individual cat, neophobia can be expressed merely as an initial bout of sniffing hesitation before eating, as a reduction in the amount eaten on the first encounter or as a complete refusal to eat the new food. The second of these possibilities can result in eventual acceptance of the new food, once the consequences of eating and digesting it have been assessed, even though the cat may seem initially reluctant to eat it. It is known from other species, and seems to be true of the cat, that the circumstances under which the new food is presented can also affect the expression of neophobia. The probability of a refusal can potentially be enhanced by almost any unusual circumstance, such as a change to an unfamiliar feeding bowl, or the presence of a strange person. Exposure to the smell of the new food alone has little or no effect on how much of it is eaten on the first occasion it is presented as a meal; one or two actual meals have to be eaten before the cat's true preference is revealed (see Fig. 6.1). To keep the new food in the cat's repertoire, it seems to be necessary for it to be presented at least once every few weeks, because it has been found that cats reacted neophobically towards a food after an interval of 3 months from the first time that they had learned that it was safe (Bradshaw, 1986).

Within their normal repertoire of foods, cats are known to appreciate variety. This is perhaps most dramatically demonstrated in young cats that have been kept on a single batch of a commercial product for many weeks, as part of a trial to confirm that product's nutritional suitability. These cats will prefer to eat almost any alternative product on the first occasion it is offered side-by-side with the diet on which they have been reared, but this effect is generally transient and the relative acceptability of the two diets tends to stabilize at the expected level (based on preferences recorded from cats that have not had a predominance of either food in their background diet) within a few days. In adult cats a similar effect can be demonstrated after only 6 days' feeding on a single product. On the other hand, such preferences for variety can disappear if the feeding environment is changed, when the cat may prefer to eat its most familiar food (Thorne, 1982). A preference for variety is not confined to house cats, nor to those deliberately kept on a monotonous diet: farm cats show a reduced preference for types of food that form a major, but not exclusive, part of their background diet (Fig. 6.2), and neophilia may even be more pronounced in farm

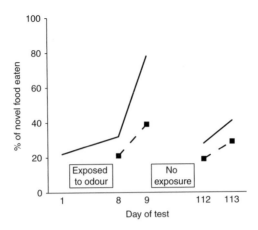

Fig. 6.1. Percentages of novel (artificially flavoured) canned cat food eaten by cats presented with choices between that food (group 1, solid lines) and an unflavoured, familiar food (group 2, broken lines). The two groups of cats were treated identically, except that only group 1 was tested on day 1. On every day between days 1 and 8 all the cats were exposed to the odour of the unfamiliar food, but were not allowed to eat it. This exposure to the odour appeared to have little effect on preference, but by the second test both groups ate more of the novel food and on the third test group 1 preferred the novel food. For the next 3 months neither group had any exposure to the novel food, after which both groups showed preferences very similar to those they had shown originally, suggesting that they were again behaving in a naïve way (data from Bradshaw, 1986).

cats than in cats fed on nutritionally complete commercial foods (Church *et al.*, 1996). Such neophilia should be adaptive in that it will discourage over-reliance on one food which may not be nutritionally adequate (Bradshaw, 2006).

Learning about toxins

The undoubted survival value of rapid learning about foods containing toxic substances applies just as much to the domestic cat as to any other animal. Stomach upsets of all kinds can alter food preferences for some considerable period (Bernstein, 1999). For example, cats can develop aversions to foods containing large amounts of sugars, because they are unable to digest these in high concentrations. In one study, exposure to a flavoured sucrose solution caused an aversion that lasted for over 7 days. In another, high concentrations of sodium chloride induced a transitory aversion to salt (Yu *et al.*, 1997). More than one unpleasant experience following a particular food may result in that food being rejected for many months. It is not known for the cat whether such experiences can also enhance the probability of neophobia, but this seems possible.

As in other animals, such aversions extend to nutritionally inadequate diets, as well as to those containing toxins. The cat's heavy requirement for thiamine,

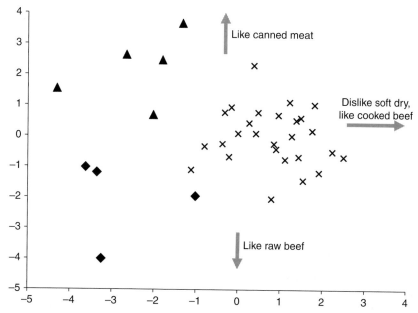

Fig. 6.2. A two-dimensional representation of the most consistent differences in food preference between cats at three farms (derived from canonical discriminant analysis of 11 preference tests). Each point represents an individual cat. The major trends in 'likings' for four of the six foods tested are indicated by the arrows. The Ducklington cats (crosses) were fed with the soft dry food by their farmer; the Wytham cats (diamonds) were given some commercial canned meat-based products by a neighbour. The Horspath group (triangles) were regularly fed on waste food from restaurants containing cooked meat, including beef, but were never given any proprietary cat foods. Each group of cats therefore shows the lowest preference for the food it had most access to (redrawn from Bradshaw *et al.*, 2000b).

particularly when it is on a carbohydrate-rich diet, has been noted above. The first symptom of a thiamine-deficient diet is often anorexia, apparently due to a rapidly learned dietary aversion (Everett, 1944). Diets that contain an imbalance of essential amino acids can produce a similar effect.

Learning about energy and nutrient balance

Cats are solitary hunters and tend only to take prey that is considerably smaller than themselves. Wolves need to be able to gorge themselves on their large, communally hunted prey while it is available, and this is probably the origin of the notoriously insatiable appetite of some types of domestic dog. A solitary hunter needs to stay at an optimum weight to be successful, and so it used to be rather unusual to find an obese house cat. The majority of cats seem to defend their set-point weight quite accurately; their appetite appears to be affected within a few hours of eating a meal that contained more or less calories than expected. Dilution with water results in accurate compensation, as does calorie enrichment with fat, so it appears that cats have accurate internal measures of the calorie contents of their foods (Fig. 6.3; see also Kane, 1989).

The physiological mechanisms underlying nutrient selection in general, and calorie regulation in particular, have scarcely been investigated in the cat, but in common with other mammals the hypothalamus plays a major role in controlling appetite. The metabolic cues that trigger changes in the hypothalamus may, however, be rather more specific and reflect the cat's carnivorous lifestyle. An inhibitor of glucostatic mechanisms, 2-deoxy-D-glucose, can cause short-term increases in food intake, as the increase in glucose levels in the blood that follows meals and inhibits continued food intake goes undetected. The normal food of cats is lower in carbohydrate than that of rats, and correspondingly it is found that cats are more sensitive to 2-deoxy-D-glucose than are rats (Jalowiec *et al.*, 1973). However, there is some evidence that low levels of sugar added to the diet are somehow not detected as calories by cats.

When offered diets with different macronutrient contents, cats reared on commercial diets have been found to select combinations of foods with reference

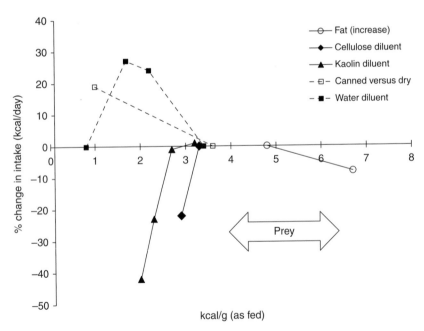

Fig. 6.3. The results of five different studies of calorie compensation by cats, expressed as percentage change in daily calorie intake compared with the standard or control treatment. Two of the studies used water as the main diluent (broken lines); three used solid materials to dilute or enrich the calorie levels (solid lines). The approximate range of calorie densities in normal prey is indicated by the arrow (redrawn from Bradshaw, 1991).

to an internal target of 52% energy from protein, 36% from fat and 12% from carbohydrate (Hewson-Hughes *et al.*, 2011). The target for protein is virtually identical to that obtained from prey (Plantinga *et al.*, 2011) while being somewhat lower for fat and higher for carbohydrate. There appears to be an upper limit of about 25% energy from carbohydrate, probably because the cat's specialized digestion and metabolism cannot process more carbohydrate than this. The mechanisms whereby cats maintain these set points are still being investigated. However, it appears that while their ancestors may have fed on nutritionally exchangeable prey items, domestic cats possess mechanisms to select from foods with a much wider range of nutrient compositions, possibly an adaptation to scavenging as an alternative to hunting.

Hunting and Predation

7

Introduction

Domestic cats are extremely efficient predators – an attribute that over the years has attracted an increasing number of investigations into how detrimental this may be to wildlife in many parts of the world. Cats are often accused of being 'ecologically surplus killers', continuing to hunt when they already have a surplus of food available, and so they also make ideal subjects for the study of the relationship between hunting and hunger. In addition, they are by far the most accessible of the small cats, and so comparisons of the typical hunting behaviour of this genus with that of other Carnivora tend to be based on the domestic cat. Leyhausen's monograph (1979) covers the latter approach in considerable detail, and only a summary will be included here.

Predatory Behaviour Patterns

Unlike the domestic dog, the domestic cat appears to have retained a complete repertoire of wild-type felid predatory behaviour. Once the cat has located its prey it will usually approach it rapidly in a crouching posture, the typical 'stalking run', making use of cover where available. On arriving within a few metres, the cat will often drop into a watching posture, in which its body is pressed flat to the ground; its forelegs are drawn back so that its forepaws are beneath its shoulders. The head is stretched forward and the ears are held erect and pointing forward. This may be followed by a second stalking run,

© J.W.S. Bradshaw, R.A. Casey and S.L. Brown 2012. *The Behaviour of the Domestic Cat*, 2nd Edn (J.W.S. Bradshaw *et al.*)

presumably because the cat judges that the prey is not sufficiently close for an effective pounce. Once within striking distance of the prey, the cat prepares to spring, chiefly by moving its hind paws further back. While in this position, the hind feet may be raised and lowered alternately, a movement seen in a much more exaggerated form in the mock pounce of kittens, when it can cause the whole of the hindquarters to wobble violently from side to side. The smaller movements of the adult, and the tail-twitching that often also occurs at this stage, appear to be maladaptive in that either could alert the intended prey to the presence of the predator. However, both sources of movement are likely to be screened from a targeted prey at ground level by the forequarters of the cat, and so have little impact on its conspicuousness. It is possible that the 'treading' movements stave off fatigue in the tensed muscles of the hind legs, fulfilling a similar function to the constant changing of the stride while galloping, described in Chapter 1. Cats watching unattainable prey, such as birds that are nearby but too high up to be worth striking at, may pull back the corners of their mouths and make an irregular chattering sound with their teeth. This has been explained as a displacement activity, but since this sound is never made in any other context it does not fall within the strict definition of a displacement activity, which should be recognizably derived from behaviour patterns seen under other circumstances (McFarland, 1985), and may have some other significance.

Cats rarely pounce on their prey, the exceptions occurring in long grass or similar cover, when the forepaws are used in a feline version of the fox's typical 'mouse pounce'. The final approach to the prey is usually a brief sprint, and whenever possible the final spring is kept short so that both hind feet are on the ground when the front paws strike the prey. If the prey moves suddenly at this point the spring may have to be corrected, and if this results in one or both hind legs leaving the ground at the moment of impact, the strike is often clumsy, possibly giving the prey the chance to escape. During the strike the whiskers (mystacials) are directed forwards, presumably to give precise information on the position of the prey, and compensating for the cat's poor vision at this close range. Prey that escapes into nearby cover, such as a crevice too small for the cat to introduce its head, may be 'fished' for with an extended forepaw.

The way that the prey is killed depends on what it is. Insects are usually pounced on with both paws, as are small birds. Small rodents are usually struck with one paw on the back or shoulder, and are then immediately bitten on the nape of the neck. If the prey is larger, say a young rabbit, or if the bite does not strike home, the cat will often hold the prey in its teeth and strike at it with its forepaws. If the prey still offers resistance the cat may fall on to its side and rake the body of the prey with its hind claws while holding it with its teeth and forepaws. The killing bite is ideally delivered to the neck, aimed so that one of the canine teeth slides between two vertebrae and severs the spinal cord. These teeth are equipped with proprioceptors, which probably assist in locating the precise target during the bite (see Chapter 1, this volume).

After the prey is dead there is often an interval before it is actually consumed. The cat will often carry the prey to cover to eat it. Although it would probably be more efficient if the prey were grasped sideways in the teeth, the normal method of transport is to drag the prey by the nape of its neck, the original target site and evidently still a strong stimulus. This method may result in the hindquarters of the prey getting dirty, and before eating it the cat will often shake the prey in its mouth to dislodge any dirt. This shaking is quite different, both in form and intensity, from the 'death shake' used by many canids to kill their prey. Cats will also shake their prey while eating it, to loosen skin from muscle, and muscle from bone. They prefer to eat mammalian prey from the head downwards, unless the skull is too strong to be broken, in which case they will discard the head and start at the neck or abdomen. Birds are usually plucked before eating; mouthfuls of feathers are torn out and then spat away, and periodically the papillae of the tongue are cleaned of debris by what appears to be grooming of the flanks, but is actually a reverse of the usual process. Further details of hunting behaviour and hunting strategies can be found in Turner and Bateson (2000).

Senses used in hunting

A great deal of the activity of kittens is directed towards learning how to hunt (see Chapter 4, this volume), and so it is not surprising that hunting technique is only very loosely based on so-called innate releasing mechanisms (see McFarland, 1985). The relative importance of the senses changes as the hunting sequence progresses from location of prey through to ingestion (Table 7.1). The primary sense used in locating prey is probably hearing – cats seem to have an inbuilt curiosity for high-pitched sounds such as rustlings, scratchings

Table 7.1. The major senses used in hunting, in the order in which they are used, starting with the initial detection of the prey and culminating in eating.

Sense	Typical stimuli	Activities elicited
Hearing	Rustling and scratching, high-frequency call-notes	Prey-seeking
Vision	Moving object of appropriate size	Stalking run, watching, springing
Touch (vibrissae)	Contact with prey	Grasping, killing bite
Touch (mouth)	Fur/feather-like surface	Killing bite (see Chapter 3, this volume)
Vision	Head/neck/body shape	Grasping, killing, picking up to carry around
Olfaction	?	Cutting open
Flavour/texture	(See Chapter 6, this volume)	Ingestion
Touch (sinus hairs)	Direction of hair on prey	Direction of cutting and eating

and ultrasonic calls. Using the latter they may even be able, with experience, to distinguish between mice and shrews, for example. Visually guided search for prey appears to rely upon the cat's appetite for the appearance of holes and crevices, which are investigated carefully. Olfaction plays a minor role, although some cats are apparently able to follow freshly laid trails of mouse urine back to occupied burrows.

Vision is used to a much greater extent in the next phase, once a visual image has been matched to a source of sound. This image is much more powerful if it is moving, and most cats require more experience before they can recognize and attack stationary prey. The killing bite is only released by furry or feathered objects, and is initially directed by the visual image of the prey's head, further guided by signals from the vibrissae (which are more heavily relied upon in the dark). The bite itself is a set of reflexes directed by trigger points around the mouth (see Chapter 3, this volume). The front-to-rear consumption of mammalian prey is mainly directed by the pile of the fur.

The development of predatory behaviour

Many of the influences on the kitten that determine the effectiveness of a cat as a predator have already been discussed (Chapter 4, this volume). To summarize, although it has been repeatedly suggested that play, particularly those aspects that involve predatory motor patterns, will affect hunting skills in adult life, the available evidence still points towards experiences involving real prey as being the most important determinants. This may be direct experience of prey items brought back to the nest when the kittens are young, and also observation of the mother's predatory behaviour. At that stage, a single experience of killing and eating a particular type of prey can have a profound effect on both prey preferences and hunting skills.

The effects of experience seem to be strongest when comparing adult cats' reactions to rats, which are much more formidable prey than the mice and small rodents that most cats are able to catch. While some cats become competent rat-killers, many show fear reactions towards rats throughout their lives. Cats that are never exposed to rats while young, or never have the opportunity to observe an adult cat killing a rat, are extremely unlikely to become rat-killers. However, not all of those that do have the right experiences overcome their initial defensive tendencies towards rats, which appear to be determined by some inherited factors that emerge in the second month of life. These factors appear to have more wide-ranging effects than just behaviour towards large prey, because cats that show this defensive attitude not only extend it to conspecifics, but are also unwilling to explore novel environments, and to purr during human social contact (Adamec *et al.*, 1983).

Motivational Aspects

The detail of the predatory behaviour of cats is especially interesting for two reasons, first because it is reputed to be disconnected from hunger, and second because it is often accompanied by what appear to be elements of play.

Hunger and hunting

Cats will sometimes start out on hunting expeditions immediately after consuming a meal provided by their owners. It is possible that this is driven, at least in part, by the cat's desire for variety within its diet, as discussed in the previous chapter, but it is also likely that diet quality affects motivation to hunt; Silva-Rodriguez *et al.* (2011) reported more predation by cats fed household scraps than by cats fed commercial food. Farmers who keep cats for rodent control often provide minimal supplementary food, in the belief that a hungry cat is a more efficient hunter, and there is some evidence from field studies that this may be so, although efficient pest control requires killing but not necessarily eating. There are at least three stages at which hunger could affect hunting behaviour: the stage of looking for prey, the kill itself and the consumption of prey. Thus a well-fed cat might not spend as much time hunting as would a hungry cat, or it might hunt less 'seriously', resulting in fewer kills, or it might kill as frequently, but consume little or none of its prey.

There is little comparative information on the amount of time spent hunting by cats with different lifestyles, but well-fed house cats may typically hunt for up to a quarter of each day, while feral (unfed) cats can spend 12 out of every 24 h searching for food (Turner and Meister, 1988). Adamec (1976) clearly demonstrated that feeding and killing are separately controlled by introducing a live rat to cats that were already feeding; almost without exception, the cats would break off from feeding, dispatch the rat, drag it alongside the feeding bowl and resume the original meal. The rat was rarely eaten, but this was predictable because in separate preference tests it was found that rat meat was less palatable than any of the foods offered. However, the most palatable food (salmon) sometimes inhibited hunting, suggesting a degree of interaction between predatory behaviour and appetite. Such an interaction is also predicted from links that have been discovered between those areas of the lateral hypothalamus concerned with killing and eating (see Chapters 3 and 6, this volume, for further discussion of the role of the hypothalamus). However, it is clear that predatory behaviour can immediately suppress feeding when prey appears; presumably the potential benefit from obtaining a second food item outweighs the risk of having the first item stolen while the cat is occupied with capturing the second.

While cats will engage in predatory behaviour whether hungry or not, the tendency to kill increases with hunger and also decreases the larger and therefore more dangerous the prey is (Biben, 1979). Prey is often not eaten immediately after the kill, and if a second prey item becomes available immediately, it may be killed in the proximity of an uneaten carcass. Thus hunger affects the initial stages of the hunt, and the probability of a kill, but visual and auditory stimuli arising from the prey itself can override any considerations of appetite, and initiate the predatory approach. Actual consumption of the prey depends on the palatability of any alternative foods as well as hunger per se.

The relationship between 'play' and 'predation'

Some people perceive cats as 'cruel' because they can be seen to 'toy' with their prey, both when it is weakened by an initial attack, but is still alive, and after it is clearly dead. It is thought that much of this type of 'play' is a displacement activity, brought about by conflicts between the need to kill and the fear of being injured by the prey. Biben (1979) found that the frequency of play was highest in two situations (Fig. 7.1). First, if hunger had reduced the tendency to kill the prey to a minimum, small and medium-sized prey were played with (hunger balanced by fear), although this was less evident for the largest prey she tested, young rats, which were often avoided completely if the cat had just been fed (fear >> hunger). If the cat had not been fed for 48 h, it would kill young

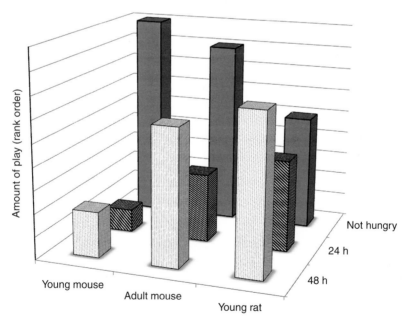

Fig. 7.1. The relative amounts of play behaviour observed in cats presented with three different types of prey, under three conditions of food deprivation. Little play was observed when the cat was hungry and confronted with prey that was easy to catch, but when it was not hungry, or when the prey was large, play formed a higher proportion of the behaviour observed (data from Biben, 1979).

mice immediately (hunger > fear), but would 'play' while and after attacking adult mice or young rats (fear and hunger both high).

Conversely, object play performed by adult cats is now considered to be, in essence, predatory behaviour performed on inanimate objects (Hall, 1998). Adult cats play more intensely and for longer the more their 'toys' resemble prey items: for example, mouse-sized toys covered with deer hide or feathers induce more play than otherwise identical toys covered with fake fur (which has a more uniform texture), which in turn produce more play than toys covered with shiny plastic. Mouse-sized toys induce more close-contact play than rat-sized toys, which would be expected if the cats were treating them as 'real' prey. The intensity of object play increases as the cat gets hungrier, again mimicking the effect of hunger on predatory behaviour, and the opposite of what would be predicted if object play were a low-priority activity (Fig. 7.2; Hall and Bradshaw, 1998).

The common observation that cats very quickly become 'bored' with particular toys can also be explained by comparison with actual predatory behaviour. The 'boredom' is actually sensory habituation – if the sensory characteristics of the toy are changed even slightly – for example, a change of colour and odour – then 'play' may resume at its original, or even enhanced, intensity (Fig. 7.3). Hall *et al.* (2002) have proposed that this habituation is a mechanism that has evolved to enhance the efficiency of predation. If a cat attacks a potential prey item and the sensory characteristics of that item do not change, then either the

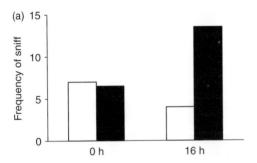

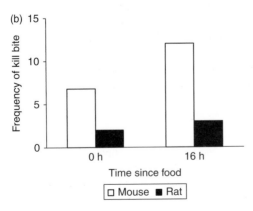

Fig. 7.2. The effects of toy size and hunger level on investigation (sniff) and predatory 'play' (kill bite). Open bars, mouse-sized toy (88 cm³); solid bars, rat-sized toy (335 cm³). (a) Mean frequency of sniff; (b) mean frequency of kill bite, in 2 min. When hungry, the cats increased their investigation of the rat-sized toy and their attempts to 'kill' the mouse-sized toy (redrawn from Hall and Bradshaw, 1998).

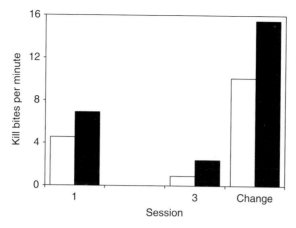

Fig. 7.3. The effect on rate of kill bite of changing the colour of a toy from black to white (or vice versa), after three 2 min sessions of play with a 5 min gap between each session. Open bars, cattery cats; solid bars, pet cats (data from Hall *et al.*, 2002).

attack has been ineffective or the item was not prey in the first place (i.e. the sensory characteristics on close manipulation are not prey-like). Toys that are manufactured to be robust thus trigger the initial stages of predatory behaviour, but their resilience to attack quickly suppresses the cat's motivation.

Hunting Methods

Cats are opportunistic predators; they take prey in approximate proportion to its availability, and frequently supplement their diets by scavenging from human refuse and carrion. Although often said to be crepuscular predators (i.e. most active at dawn and dusk), cats are in fact very flexible in timing their hunting sorties, usually to coincide with the main activity periods of the most readily available prey. House cats appear to time their expeditions around periods when food is unavailable from their owners. Very high or very low temperatures are not favoured, so that in very hot weather cats will tend to hunt at night, and the reverse in winter.

The hunting methods of cats can be classified according to the strategy adopted, which will depend loosely upon the type of prey being sought. A sit-and-wait strategy may be adopted where there is a concentration of burrows containing suitable prey. On the other hand, small birds need to be stalked. Leyhausen (1979) considers that the domestic cat is generally a rodent special-ist, because its preferred sit-and-wait strategy is much better suited to catching mice or young rabbits straying from the mouths of their burrows than it is for a bird that can move in any direction, including upwards. None the less, some individual cats do seem to become specialist bird-hunters. Colonies of burrow-nesting birds can be raided by cats going down the burrows themselves if they are wide enough.

Cats are reasonably successful predators, since between 40 and 65% of free-ranging cats have identifiable prey in their stomachs (Turner and Meister, 1988). While the availability of prey has a large effect on what is taken, larger prey is often captured at a lower rate (measured as pounces per catch). Despite this, the larger size of rabbits means that they are actually more rewarding (measured as calories per pounce) than small rodents. An optimizing predatory cat should therefore take more rabbits than rodents (if both are equally available), and this does seem to be true of feral cats that depend on hunting. House cats, on the other hand, take fewer rabbits than optimization would predict, presumably because at their lower levels of hunger, fear overcomes the urge to kill larger prey.

Turner and Meister (1988) have also been able to show that queens with kittens to feed are much more efficient predators than other cats. They followed 143 hunting expeditions made by 23 farm cats, which were provided only with bread and milk and table scraps by their owners. The mother cats caught prey on average once every 1.6 h, while the others were only successful just over twice every 24 h. The average duration of hunting excursions, about 30 min, was the same in both groups, but the mothers travelled twice as fast. This did not reduce the number of potential prey-containing sites that they discovered, which they therefore located at twice the rate attained by the others. They spent only about 1 min investigating each location before moving on to the next, compared with about 3 min for the non-mothers, and were more successful at actual capture of prey (3.4 pounces per capture for mothers, 12.3 for non-mothers). The presence of kittens in the nest stimulates the bringing home of prey, possibly via the hormone folliculin. When the kittens are newly weaned the prey is killed before being transported, but this is replaced later by disabled but live prey (see Chapter 4, this volume). Cats without kittens, including intact toms, will also occasionally bring prey back to their primary 'home', but the reason for this is not known.

Ecological Aspects of Predation

Wildlife biologists are frequently critical of the damage that cats may do to local populations of small mammals, reptiles and small birds. There is no doubt that on many species-poor oceanic islands, the introduction of feral cats has had devastating effects on ground-nesting birds and other species that lack the behavioural, morphological or ecological adaptations to resist mammalian predators. Most studies of predation in the last 20–30 years have focused on the types of prey taken by feral and domestic cats and the effects of this predation on prey species. A considerable body of data is now available on this subject and has been reviewed by various authors (Nogales *et al.*, 2004; Bonnaud *et al.*, 2011; Medina *et al.*, 2011). By recording patterns of hunting behaviour and habitat use from such studies, ecologists have aimed to identify effective ways of controlling feral populations. More controversially, possible methods for controlling or restricting movement have been investigated even for populations of cats that are largely owned.

Species taken

Much of the information available on the prey taken by cats actually refers to what is consumed, as ascertained by analysis of faeces and gut samples, and generally omits prey items that are caught and not eaten. Exceptions are reports of prey brought home by house cats, and the uneaten remains of prey found in the field in those rare situations where cat kills can be confidently distinguished from the victims of other predators, such as on oceanic islands.

Mammalian remains are commoner than remains of birds, and mammals also form a higher proportion of vertebrate prey in studies of prey brought home, thus confirming the idea that cats are specialized predators of small mammals. This seems to be the case in all regions of the world where it has been studied. Reptiles can be important prey at low latitudes, but are rarely a significant part of the diet at latitudes above 35°. Invertebrates, including insects, spiders, isopods, crustaceans and molluscs, are often recorded, occasionally in large numbers, but their generally small size means that they are unlikely to make a major contribution to overall energy intake, except perhaps in juvenile cats that are not sufficiently mature to catch many mammals.

Among the mammals, the main species taken vary between the continents. In North America and Europe, ground-dwelling rodents are popular catches, particularly the common vole (*Microtus arvalis*) and the field vole (*Microtus agrestis*). Young rabbits and hares are also common, and because they are larger than voles can make the major contribution towards weight of food eaten. Murid rodents, including mice (*Mus musculus*) and young rats (*Rattus* spp.), are less commonly eaten than voles, possibly because they are less palatable, but house cats frequently bring these back from the field uneaten, so they are probably often killed but not eaten. Certainly many farm cats would not earn their keep if they restricted their hunting to rabbits and voles. Shrews (*Sorex* and *Talpa* spp.) are certainly unpalatable and are rarely eaten. In North America, but not in Europe, members of the squirrel family (Sciuridae) are also common prey, including ground squirrels and chipmunks. In Australasia introduced rabbits, rats and mice are supplemented by marsupials and other native mammals of similar size.

The prey taken on islands depends very much on what is available. In a review of 72 studies of the diet of feral cats on islands, Bonnaud *et al.* (2011) reported that cats fed on a total of at least 248 different species ranging from large birds and medium-sized mammals to small insects. As well as the cats, mice and rats and rabbits have also often been introduced, and form part of the diet, but in contrast to the diet on the mainland, birds, mainly seabirds, become much more important on islands. These can include petrels, terns, noddies and penguins, depending on latitude and the particular characteristics of the island.

Effects on prey populations

Cats probably do more damage than anywhere else where they have been introduced to oceanic islands. Perhaps the most extreme example of this is the short history of the extinct Stephens Island wren, all 15 specimens of which were brought home by the lighthouse keeper's cat in 1894. More recently a single cat was responsible for the extinction of a subspecies of the deer mouse (*Peromyscus guardia*; Vazquez-Dominguez *et al.*, 2004). Domestic cats are dietary generalists and feed on many types of native and introduced prey – this ability to consume such a variety of different prey type increases their potential to impact on virtually any endemic species lacking the necessary adaptations to avoid capture. Where exotic prey such as mice, rats and rabbits have also been introduced it has been shown in some cases to compound the problem by enabling a growth in the predator population. This subsequently has a greater impact on the more vulnerable native prey (e.g. Macquarie Island, where a species of parakeet coexisted with cats until rabbits were introduced; see Medina *et al.*, 2011).

Ground-feeding and flightless species are particularly vulnerable, as are small species of seabirds that nest on the surface and larger species of seabirds whose burrows are large enough to be entered by cats. Local extinctions of endemic mammals and reptiles are also well documented. Impacts of feral cats on native island vertebrates have been reported from at least 120 different islands on at least 175 vertebrate species (25 reptiles, 123 birds and 27 mammals: see review by Medina *et al.*, 2011). They also calculated that globally cats have contributed to at least 14% of the bird, mammal and reptile extinctions on islands and are the principal threat to 8% of critically endangered birds, mammals and reptiles. The domestic cat is also on the list of the 100 worst invasive species (Nogales *et al.*, 2004).

On the mainland it is generally more difficult to assess the precise impact of cats because there are usually a whole range of predators present, and prey numbers can more easily recover by immigration than on islands. Feral cats have certainly been a major contributor to the decline in numbers of native fauna in some mainland areas (e.g. the Eastern barred bandicoot (*Perameles gunnii*) in Australia: see Dickman, 2009). Occasionally they may have a positive effect on certain species. Dickman (2009) describes how an inverse relationship was observed between feral cat activity and predation on artificial tree-nests in Sydney, Australia. This suggests that, through their own hunting, cats may have reduced the impact of nest predators such as rats and possums on tree-nesting birds. However cat activity in this study also correlated negatively with overall bird species richness, so cats may have benefitted tree-nesting species at the expense of species that nest closer to the ground. The ecological impact of pet cats remains controversial. Pet cats living as well-fed house pets and receiving all their food from their owners will still hunt opportunistically. They undoubtedly take large numbers of prey but most of the species killed are small, short-lived and have high reproductive rates. There is also a distinct skewed frequency in the number of prey brought home by different cats, with many studies finding that just a small fraction of the cat population brings home the majority of prey (Tscanz *et al.*, 2011). Thus the actual impact on prey populations depends on the degree to which cat predation is additive or compensatory (for example, taking young or weak individuals) compared with other sources of mortality (Tscanz *et al.*, 2011). Studies on the diets of animals at different ends of the management spectrum of feral versus well-fed owned pets show that cats that receive less food from humans hunt more. Feral cat scats usually contain more than one prey individual per scat, whereas studies of owned, well-fed cats report predation rates of mostly less than five vertebrates per cat per month (Silva-Rodriguez and Sieving, 2011).

The sex of the cat does not appear to affect the amount of prey caught. Older cats generally catch fewer prey than do younger cats, although Gordon *et al.* (2010) found older cats in their study (6 years plus) were more likely to catch rats than cats aged 2–4 years. There may also be an interactive effect of sex and age on overall prey composition (e.g. Gordon *et al.*, 2010, found that female cats caught the most mice at an earlier age (2–4 years) than did male cats (4–6 years)).

Some argue that the impact of predation by domestic cats is exaggerated and that because cats are successful at controlling pests, native species may benefit from reduced rodent populations. Other authors maintain that overall predation rates outweigh the benefits of any rodent catching.

Control measures to reduce predation by cats

Many recent studies have focused on identifying ways of reducing predation by cats, both ferals and domestic pets. Where feral cats have decimated island populations some ambitious conservation methods have been tried, for example transfer of the kakapo, a large, flightless, endemic parrot in New Zealand to cat-free islands. On many islands, however, conservation has meant a programme of complete eradication of feral cats. In their review of such eradications, Nogales *et al.* (2004) reported that globally feral cats have been removed from at least 48 islands. The majority of these islands are small (less than 5 km^2), the largest successful eradication campaign being on Marion Island (290 km^2). In some places these eradications have resulted in dramatic recoveries of threatened vertebrates (e.g. the bird *Philesturnus carunculatus* on Little Barrier Island, New Zealand: see Medina *et al.*, 2011).

Reduction of predation by domestic pet cats has also been investigated in some detail (e.g. Gordon *et al.*, 2010). Measures to reduce predation opportunities include curfews (keeping cats inside at night) for domestic cats, the creation of cat-free buffer zones around vulnerable areas and, in places, the introduction of legislation to encourage responsible cat ownership. The effectiveness of these can be difficult to quantify; in some places where controls have been introduced species diversity and abundance of prey appear to be more affected by vegetation characteristics than by restrictions on the movements of pet cats (Lilith *et al.*, 2010). Curfews, whilst providing some protection for nocturnal mammals, will have little effect on predation on birds and reptiles that are mostly diurnal.

Reducing the efficiency of hunting by cats has also been investigated by attaching various devices to them such as bells, collar-worn 'pounce protectors' and sonic alarms. Bells are probably the most popular choice of deterrent device by owners since they are less restrictive to the cat. Results of studies on these have proved equivocal, reducing the catch by half in some cases and apparently not at all in others. This may be due in part to different study methodologies (see Gordon *et al.*, 2010). It has also been suggested, although not proved, that some prey species may not necessarily associate the sound of a bell with danger. Although it is possible that proficient hunters may adapt their technique when wearing bells and learn to stalk their prey silently, there seems to be no evidence for this in the short-term studies that have been conducted (Ruxton *et al.*, 2002; Nelson *et al.*, 2005).

Effects of food supply on cat densities and home ranges

Cats exploit a very wide range of habitats, and their densities can range between less than one to more than 2000 cats/km^2. Liberg *et al.* (2000) hypothesize that the density of both free-ranging house and feral cats is determined ultimately by food abundance. Using data from 28 different studies, they classified populations into three different density categories based on their method of food provision:

1. Density >100 cats/km^2: urban environments with rich clumped food supplies from a variable mixture of garbage, fish dumps and direct provisioning.
2. Density 5–50 cats/km^2: farm cats, or feral cats feeding on more thinly clumped food resources such as seabird colonies.
3. Density <5 cats/km^2: rural feral cats feeding on widely dispersed prey such as rodents and rabbits.

For female cats, home range size as well as density seems to be determined largely by the availability of food, even though these sizes have been found to vary from 0.1 ha in a Japanese fishing village to 170 ha in the Australian bush. Density and home range size appear to be negatively correlated – in other words, where densities are high around a clumped resource, home ranges are correspondingly small as the cats do not need to travel so far to find food. Food distribution also has an effect – ranges tend to be larger than predicted when food is more patchily distributed (see Liberg *et al.*, 2000). Farm cats that are well provisioned at home also have larger home ranges than predicted, probably because they have to travel between their resting sites in the farm and their hunting grounds. Thus availability of shelter can also interact with food availability to affect spacing patterns.

Considerable overlap of home ranges occurs when females live together in a group. The social structure of these groups will be described in detail in the following chapter, but the primary reason for their formation is the existence of a large and predictable food supply, which it would not pay any one individual to defend. This is because domestic cats rarely if ever hunt in groups, and if they do (as, for example, observed of a brother and sister by Turner and Meister, 1988) there is no evidence that this enables them to take prey any larger than they would normally hunt when solitary.

The food source is almost invariably of human origin, including garbage dumps, concentrations of rodents on farms or direct provisioning (and occasionally two or all three at once). There is little interchange of females between these groups, so it is usually assumed that each group defends its resources against intruders from other groups. The home ranges of such females overlap considerably, particularly around the major food source, although each may have a non-overlapping hunting ground. Where several food sources exist, more complex patterns of overlap may develop within a population. In one study (Say and Pontier, 2004), feral cats living in parkland around an urban hospital were fed at five different feeding sites. Each female generally used only one of these five sites and average home range overlap between females using the same feeding site was up to 83%. Percentage home range overlap between females utilizing adjacent feeding sites was nearer 45%.

The home ranges of intact adult males are usually much larger than those of adult females (Fig. 8.1); if the home range was determined entirely by considerations of food supply, the average ratio of 3.5:1 male:female home ranges would predict a 5.3:1 difference in weight between the sexes, whereas males are rarely more than 50% heavier than females. The males in Say and Pontier's (2004) study included up to three of the five available feeding sites in their home range, despite just one site being sufficient to meet their energy requirements. This supports the idea that male spacing patterns are heavily influenced by access to females rather than just food sources (Liberg *et al.*, 2000; Say and Pontier, 2004; see Chapter 8, this volume, for further discussion).

Neutered males often use reduced-size home ranges that are not significantly larger than those of the females in the same area (e.g. Morgan *et al.*, 2009). Presumably, for neutered cats, removal of the urge to search for a mate allows them to roam less and remain closer to their source of food, whether that is their owner's home – in the case of the suburban house cat – or a communal provisioning site in the case of more densely populated neutered feral populations.

For house cats sharing a home and therefore a food source, there will of necessity be some overlapping of home range, and this has been found to be more pronounced if the cats are related (Barratt, 1997). This is discussed in more detail in Chapter 9, and the relationship between access to resources and undesired behaviours in Chapter 11. Time-sharing (the communicative aspects of which are discussed in Chapter 5) may enable some less compatible cats to have ranges that overlap spatially but not temporally.

Social Behaviour

8

Introduction

For many years it was thought that the domestic cat was an essentially solitary creature, which only tolerated the close proximity of its conspecifics for mating and while rearing offspring. It has often been said that the lion is the only fully social felid. However, while it is true that the social system of the lion is complex, involving communal hunting and cooperation within both male and female groups, social interaction between members of otherwise solitary felid species has been observed (Caro, 1989). Studies of the social structure of groups of domestic cats have shown that they are not just artefacts of the conditions under which house cats are kept; in fact, social structure is most clearly present in groups that are barely tolerant of human company. Looking objectively at the vast range of population densities recorded for domestic cats (Fig. 8.1), it seems very unlikely that a uniform system for intraspecific interactions could be effective, when individual cats can find themselves spaced at anything from an average of ten to several thousand metres apart. In common with other members of the Carnivora that can adapt to a wide range of population densities, the social structure that pertains to each population of domestic cats varies according to the ecological circumstances in which they find themselves (Macdonald, 1983). Essentially, groups may be formed when the availability and dispersion of food allows two or more individuals to live in close proximity, and on all the occasions that this has been documented much of this food has stemmed from man's activities. This inevitably raises two questions: whether such coalitions can ever occur without man's tacit collaboration and, if not, how the necessary behaviour patterns evolved, unless they are a by-product of domestication.

© J.W.S. Bradshaw, R.A. Casey and S.L. Brown 2012. *The Behaviour of the Domestic Cat*, 2nd Edn (J.W.S. Bradshaw *et al.*)

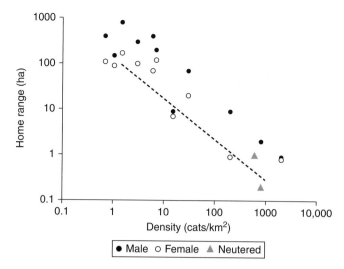

Fig. 8.1. The relationship between home range size and cat density, for entire males, females and neuters. The broken line indicates the size of home range expected if each part of the available space was allocated to the home range of one male and one female; points well above this line indicate overlap between the ranges of members of the same sex, while points well below indicate that not all the available space is used (data from Liberg and Sandell, 1988, with additions from Chipman, 1990, and J.W.S. Bradshaw, unpublished).

Solitary Cats

The question of whether cats kept singly in households are truly solitary will be left to the following chapter, but it can be argued that, like the dog that appears to perceive its human owners as part of its 'family', most cats direct species-typical behaviour towards their human keepers. Truly solitary cats that have little conspecific or human contact for much of the year have not been studied in much detail, not least because they are difficult to locate and harder to approach. The European wildcat (*Felis s. silvestris*), although not the closest wild ancestor to the domestic cat, is thought to be almost entirely solitary and may be genetically predisposed to be so, since even its kittens are very difficult to domesticate. Largely solitary populations of *Felis s. catus* are also known, for example those in the bush of south-east Australia and on some uninhabited islands. Generally, these populations support themselves by hunting and, because it is rare for prey suitable for domestic cats to be highly abundant in any one location for any length of time, social groups are rare. Under such circumstances cats are rarely seen in the company of other cats, except for male/female pairs at oestrus, and females with juveniles. The adults are usually territorial to some extent, although the mechanism whereby such territories are maintained, given that individuals so rarely encounter one another, is not clear.

When food is more patchily distributed, but each patch or group of patches is still insufficient to support more than one cat, home ranges can overlap quite extensively, but some system of temporal separation may then operate, so that two cats rarely hunt the same area at the same time. Scent marking has been implicated in maintaining this 'time-sharing' arrangement (see Chapter 5, this volume). As discussed in Chapter 7, the home ranges of females encompass sufficient food and shelter for their needs and those of their offspring while they are dependent. Even when cats are well dispersed, the home ranges of males are larger than those of females (Fig. 8.1) – 3.5 times larger on average. Male ranges of up to 10 km^2 have been recorded, and it is likely that some males, labelled as 'transient' in most studies, either have even larger home ranges than this or are more or less nomadic (Liberg and Sandell, 1988). Male ranges appear to be dictated by the availability of breeding females, whether these are solitary or social, and the degree of competition for them; the factors determining the size of individual male ranges will be discussed further below.

Group-living Cats

Both wide-ranging surveys and more detailed studies of small areas have documented the existence of colonies of domestic cats. The conditions for the establishment of these colonies almost always involve a localized concentration of food, arising deliberately or accidentally from human activities (Kerby and Macdonald, 1988). Some occur around rubbish dumps, studied in locations as diverse as Portsmouth naval dockyard in the UK and a Japanese fishing village. Others are more direct products of provisioning, such as the semi-wild populations often found on industrial and hospital sites or in various public areas in Rome. Farm cats are provisioned both directly and indirectly, by direct handouts of food, by the concentration of rodents in grain stores and sometimes by the theft of food intended for livestock. The size of the colonies seems to be very variable; in one survey of 300 colonies on industrial sites in the UK, most comprised between one and ten individuals but 7% contained over 50 cats. The critical factors determining the size of the colonies are the availability of food, infant mortality due to feline panleucopenia and other viruses, and direct killing of adults by man.

Small colonies may consist of a single social unit, while larger colonies usually contain several such groups. Most of the groups that have been studied have turned out to consist of females, usually related, together with their offspring, including immature males, and unrelated mature males. Spontaneous movement of females between groups seems to be rare, probably because while females within a group are generally tolerant of one another, they usually attack outsiders on sight, both males and females, and these attacks generally become more intense when there are young kittens in the group. Mature males are only loosely attached to any one group and, as with solitary cats, their home ranges tend to be larger than those of females.

Sexual Behaviour of Females

Apart from the contact between mother and offspring, described in Chapter 4, this volume, the only essential component of social behaviour required of a solitary female is that leading up to mating. In solitary individuals, there will be a strong territorially based tendency to attack any cat, and one of the main functions of courtship behaviour may simply be to bring the sexes together without fighting for long enough for copulation to take place. Even within a colony of cats, where all individuals are familiar to one another, if a male shows more than a fleeting sexual interest in an anoestrous female, the female will move away and if necessary spit and strike out with her claws. In pro-oestrus the behaviour of the female changes, first subtly as a tendency to move about more than usual, and then as an increase in object-rubbing (see Chapter 5, this volume). Males that approach at this stage are greeted with less hostility than before, but prolonged contact is still not tolerated. Over the next 24 h the rubbing increases in intensity and persistent bouts of rolling occur, accompanied by purring, stretching and rhythmic opening and closing of the claws. Males are now permitted close to the female, and may be allowed to lick her, but at this stage any attempts at mounting result in a considerable display of aggression.

Complete sexual receptivity does not ensue until the beginning of oestrus, which is often indicated by an abrupt change in behaviour. The rolling of pro-oestrus is interrupted by the female adopting the lordosis position, suddenly crouching with her head close to the ground and her hind legs treading and partly extended. Her tail is laterally displaced, uncovering the perineum, and it is at the moment that this display first appears that an experienced male will first attempt to mount. Grasping her neck in his jaws, he begins his copulatory thrusts, while the female treads backwards with her hind legs so that the perineum is rotated further backwards and upwards, until the male achieves intromission. At this point the female usually emits a loud, piercing cry, and within a few seconds jumps away from the male and turns on him, spitting and scratching. The female then grooms her genital region and begins to roll vigorously. Several minutes later she will adopt the lordosis position again, and this cycle of events can be repeated, with the interval between copulations lengthening, over the next 1 or 2 days (Michael, 1961). Multiple copulations are normally needed to trigger ovulation, and without copulation ovulation does not occur. On the one hand, it has been suggested that this induced ovulation is an adaptation to solitary living, preventing the female from ovulating wastefully, before she has been able to attract a male. On the other hand, the whole process of pro-oestrus, when the female is attractive to males but not receptive, and also the need for multiple matings, could also be devices to enhance competition between several males courting one female, and may therefore not only be an adaptation that enhances the female's fitness, but also to living at high density.

Social aspects of maternal behaviour

Females tend to stay within a single social group for much of their lives; a solitary cat may occasionally join an established group, and formerly group-living females can become solitary, but migration between groups by females seems to be rather rare (Liberg and Sandell, 1988). Devillard *et al.* (2003) found that some females dispersed at 1–2 years of age from their study colony of intact, high-density cats and suggest that cats may show intraspecific variability of dispersal patterns related to changes in environment, density, mating system and social organization. Their general tendency to remain in their natal group, combined with the amicable treatment of fellow members of the group, contrasting with aggression towards outsiders, implies that group-living females defend a communal core territory, which is likely to include their denning sites and their major source of food. Large colonies usually consist of one or more 'central' groups of related females that defend areas that contain the best resources for feeding and denning, together with other 'peripheral' females that are either solitary or form smaller groups that occupy inferior territories (Macdonald *et al.*, 2000). Central females generally produce more surviving offspring than peripheral females, suggesting that combining their efforts to defend their resources gives them a genuine reproductive advantage.

The other obvious benefit from cooperation between the female members of a group is the communal denning and nursing of kittens. While in large groups such collaborations tend to be within mother–daughter pairs, in small groups all the adult females may nurse each other's offspring and the litters are often pooled in communal dens. Females will aid in the birth, then groom, nurse and guard each other's kittens. Pooling of litters may also facilitate the learning of a family identity among the kittens, enabling them to form kin-based alliances once adult. Other advantages may also exist for such pooled litters. In one study, communal litters occupied twice as many nests as solitary litters over the first 6 weeks after birth. The reasons for such frequent nest moves are uncertain, but may help to maintain concealment of kittens from predators and avoid possible infanticide (Feldman, 1993). Kittens are gradually moved to nests nearer to the food source as they develop towards independence in both solitary and communal litters, so facilitation of weaning may be a reason for repeated moves. Whatever the reason, kittens from communal nests spend less time alone and are therefore at less risk from predators during such movements than kittens from a female raising her kittens alone. Kittens from pooled litters also leave the nest earlier than those in solitary litters, suggesting that communal denning may confer some developmental advantages on the kittens.

However, there are also disadvantages to communal living. Several contagious diseases are ideally suited to transmission within a communal den and between cats sharing concentrated food sources. On the oceanic Marion Island, the feline panleucopenia virus was introduced deliberately as a measure to control the feral cat population, and not only was there a fourfold decrease in the population, but the proportion of cats living in groups declined also. Once a virus infects a farm cat colony, it may spread rapidly to all members, and not just within core groups that share nests and resting places (Macdonald *et al.*, 2000): transmission between colonies may simply be low because migration of cats from one farm colony to another is rare, as is close contact between cats in different colonies.

Male Behaviour

There is very little evidence for cooperative behaviour between intact male cats; for example, Dards (1983) never observed any amicable behaviour between any of the mature males in a Portsmouth dockyard. When two unfamiliar males meet for the first time they may initially sniff each other, but this quickly gives way to aggression, including the arched-back posture (Fig. 5.1), growling and yowling. If one male goes on to the defensive, indicating the other's superiority, it will tend to crouch, hiss and strike out with its fore-claws. After repeated encounters, overt aggression is reduced but the original winner will tend to spray urine and rub objects more frequently than the loser (de Boer, 1977b). Affiliative behaviour between intact males appears to be restricted to juveniles from the same family group, which may associate together until they become sexually mature and disperse (Macdonald *et al.*, 2000).

Territoriality and the mating system

The size and organization of home ranges and the mating system of male domestic cats tend to vary according to the environmental conditions and density of the particular cat population. The home ranges of males are, on average, about 3.5 times larger than those of females living under the same ecological conditions, but this apparently simple relationship conceals a great deal of variability. In rural areas where cats are living at low density (100–300 cats/km^2; Liberg *et al.*, 2000), males maintain exclusive home ranges which may overlap those of several females. They may defend access to females in oestrus during the mating season thereby securing most, if not all, of the matings. This system is classified as polygynous (Pontier and Natoli, 1996; Say *et al.*, 1999), whereby each male copulates with several females at each oestrous period. The females may interact at a high rate with such a male when he is present, and may appear to be trying to prolong the male's presence in that part of his home range that encompasses their core area (Macdonald *et al.*, 1987). Outside the breeding season, male ranges in such low-density populations tend not to overlap a great deal, but in the breeding season their ranges expand and overlap considerably, as they try to gain access to as many receptive females as possible. This may be exaggerated when the intervention of man reduces the number of mating opportunities through the neutering of females within a male's normal home range. Yearling males in this environment often stay close to their mother or maternal group, but as they grow they come under increasing attack from older males. In their second or third years, they usually disperse away from their mother's home range. Individual males that do not emigrate appear not to become sexually mature. At this stage they may become strongly solitary, avoiding contact with all other cats, but most eventually come to challenge the successful breeding males for access to females. Males in rural populations may not start reproducing until around the age of 3 years (Say *et al.*, 1999).

Where cat densities are high (up to 3000 cats/km²; Liberg, 2000), such as in urban populations, territories are indistinct and overlapping, encompassing large multi-male–multi-female groups of cats (Natoli *et al.*, 2000). Here, even those males holding territories may be unable to monopolize all of the receptive females, leading to a promiscuous mating system in which both males and females mate with several partners at each oestrous period (Natoli and De Vito, 1991; Say *et al.*, 2002). Aggregations of males around receptive females are commonplace where cat population densities are high, and their most surprising feature is that under these circumstances the males are often less aggressive towards each other than when there is no available female in the vicinity. Natoli *et al.* (2007) point out that the cost associated with the exclusion of all other males would be too high for the resident male – while he was fighting with one competitor, others might exploit the situation and mate successfully with the female. So he opts instead for sharing the mating opportunities and therefore the paternity of litters with other males (Say *et al.*, 1999, 2001). Males in dense populations begin reproducing as soon as they reach sexual maturity (Say *et al.*, 1999) and may be less likely to disperse. Devillard *et al.* (2003) observed no evidence of male dispersal at any age over 8 years in the colony they studied. Possibly the reduced inter-male aggression seen in high-density populations enables juvenile males to remain.

Genetic analysis of paternity in Devillard *et al.*'s (2003) study showed that highly inbred mating occurred in that colony, presumably due in some part to the non-dispersal of males. There did not appear to be any decrease in litter size, survival probabilities or body weight of offspring as a consequence, suggesting that perhaps there are only low costs associated with inbreeding in domestic cats (Devillard *et al.*, 2003). Where colonies are not isolated, there may be more outbreeding than inbreeding (Yamane, 1998).

Reproductive success in males

Studies of the mating system have advanced considerably since Menotti-Raymond and O'Brien (1995) developed a technique using DNA to determine the paternity of kittens. Researchers are now able to accurately measure male reproductive success under different social and environmental circumstances.

Say *et al.* (1999) analysed the paternity of kittens born in two contrasting habitats (rural versus urban) with different densities of cats (234/km² versus 2091/km²). They found a high rate of multiple paternity in the urban population, with 70–83% of litters having more than one father, whereas this was much lower in the rural population, with only 0–22% of litters having more than one father (Fig. 8.2). One might expect the rural males to obtain exclusive siring of kittens; however 'satellite' males are sometimes able to copulate with females in the absence of the resident male (Say *et al.*, 2002).

Males having a territory overlapping that of females, therefore, have varied reproductive success depending on the local population density. In urban areas, such males do not have complete control of receptive females or full paternity of single litters. However, they do achieve the highest reproductive success: Natoli *et al.* (2007) found that the resident male sired the highest percentage of kittens but monopolized only one whole litter out of nine, and co-sired the highest number of other litters.

Males with the largest home ranges include the most female home ranges and have the highest reproductive success (Say and Pontier, 2004). They may also successfully reproduce with females whose home ranges do not overlap theirs – in Say and Pontier's (2004) study, 28% of kittens were sired by males whose recorded home ranges did not overlap with the kittens' mothers' home ranges. This indicates that males may make quick excursions outside their ranges to find new mating opportunities. Thus, although ranging more widely decreases the probability of siring all the kittens in any single litter (Say *et al.*, 2001), apparently using a large area increases fertilization success by allowing a male more mating opportunities (Sandell, 1989). In group-living situations, therefore, a male's ability to maintain a large home range may be one of the main keys to improved mating success and, interestingly, has been found to be independent of his agonistic behaviour, which is conventionally interpreted as an indicator of 'social rank' (Say and Pontier, 2004).

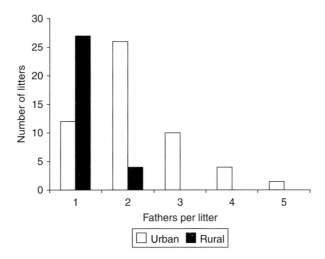

Fig. 8.2. Distribution of the number of fathers per litter in a rural and an urban population (redrawn from Say *et al.*, 1999).

Body weight has also been shown to affect reproductive success. Yamane (1998), studying a group-living population on a small island, identified two different reproductive tactics according to body weight of the male. Heavier males courted females both of their own group and of other groups. Lighter-weight males courted females only of their own group. Observations of the extra-group courting attempts suggested this may be a fairly unsuccessful tactic by the heavy males as their courtship rank was reduced in such groups. DNA evidence, however, revealed that over half the kittens from groups of females were fathered by 'extra-group' males. Yamane suggests that the discrepancy between observed copulation and actual paternity may indicate that some female choice may be operating on the mating system in that particular group of cats.

The question of whether females exert choice over which males sire their kittens has produced equivocal results. Natoli *et al.* (2000) found that, although females copulated with several males during a single oestrous period, there was no evidence of mate choice to be found. Ishida *et al.* (2001), however, found that their females did not accept all mounting or copulation attempts and in particular appeared to avoid inbreeding with close kin (1/4 or higher degrees of relatedness). More distant relatives were not refused.

One other important variable affecting the reproductive success of males is the degree of oestrus synchronization amongst females. When oestrus is asynchronous a more competitive male can attain higher reproductive success than lower-ranking males through priority of access to females. This allows him to copulate with more females or to copulate with the same female several times to ensure paternity (Say *et al.*, 2001; Say and Pontier, 2004). When oestrus is synchronized, as is often the case in dense urban populations, a single male will not be able to monopolize all the females and other males will be able to mate with them too. Say *et al.* (2001) found that variance in male reproductive success was four times greater in those years when females bred asynchronously, with dominant males siring the highest proportion of offspring.

Infanticide

When one coalition of male lions succeeds in ousting another group of males from a group of females, they usually kill all the cubs in the pride, thereby bringing the lionesses into oestrus more quickly than if they had completed lactation through natural weaning. Examples of infanticide by male domestic cats have been reported (e.g. Macdonald *et al.*, 1987; Pontier and Natoli, 1999). In the latter study, six cases of infanticide were directly observed in rural populations of cats. It is unclear just how common this phenomenon is, although it may be one factor causing the aggression shown by nursing females towards strange males. It has not been described for high-density populations, and Say *et al.* (2001) suggest that synchronicity of oestrus in females in densely populated breeding groups may help reduce the risk of infanticide through the mixing of paternity that ensues (see section on reproductive success in males). Since the birth interval in the domestic cat is only 4 months, compared with 19 months in the lion, the advantage to be gained from infanticide in bringing a female into breeding condition may not be great (Natoli, 1990). In temperate climates the second (autumn) litter rarely produces surviving offspring, but it is possible that its chances of success are increased by bringing the time of conception forward by even a few weeks. Infanticide might also be a successful strategy when cat density is low and adult females, food or suitable nest sites are at a premium. A non-breeding male could enhance the chances of his own future offspring surviving by attempting to kill all the offspring of the current resident male, so that when that male lost his position, and he took over, his own kittens might be born in the best nest sites (as selected by the surviving females) and be fed sufficiently well to reach adulthood. However, it is still unclear whether infanticide is a common strategy among feral cats, or an aberration only practised by a very few. This is discussed in more detail by Pontier and Natoli (1999).

Social Communication

We have seen that, where there is a sufficient concentration of food, cats form more or less stable groups, the basis of which is usually the cooperative rearing of kittens by related females. The role of males in these groups is still poorly understood, and may vary with population density. The membership of these groups is generally stable outside as well as during the breeding season, and as for other social Carnivora, the social structure is maintained by a 'glue' of interactive behaviour patterns. In other species these patterns have been interpreted as indicating some kind of hierarchical organization, and an individual's position in the hierarchy is usually estimated by the behaviour patterns it exhibits towards other members of its group. Some of the most revealing of these patterns are those indicating submission, where one individual signals that it wishes to 'back down' from an encounter. However, no such pattern has been identified in the cat, which tends to ward off aggressive approaches with defensive, rather than submissive, behaviour. Whereas consistent patterns of aggressive interaction have sometimes been identified between males, and interpreted as a 'dominance hierarchy', the collaboration between females is more cooperative than hierarchical, because it is common for all the females in a group to breed simultaneously. Certain behaviour patterns appear to strengthen the bonds between individuals, and to build up a group identity. These social behaviours include scent marking, mutual grooming (allogrooming) and mutual rubbing (allorubbing), and these will be discussed in turn. To put these into a social context, the flow of interactions within social groups will be described first.

In a number of published studies, not only has the behaviour of particular individuals been recorded, but also the partners they chose to interact with, whether those individuals were considered separately, as close relatives or as representatives of a particular age/sex class. In colonies that obtain some of their food by hunting, the cats tend to visit their core area independently, neither avoiding one another nor tending to always be present with a particular partner. However, when they are present they tend to single out particular partners for interaction. For example, Wolfe (see Curtis *et al.*, 2003) studying two colonies found that some adult cat dyads were within 1 m of each other more often than would be expected by chance – these he referred to as preferred associates. Proximity was not dependent on location, so cats were not simply aggregating at preferred resources at the same time. Some cats appeared actively to avoid one another too – in a colony of intact cats these were mainly male–male pairs, while in a neutered colony there was no effect of gender (Wolfe, 2001; see also Crowell-Davis *et al.*, 2004). Macdonald *et al.* (2000) discuss in more detail the preferences cats had for proximity to both kin and particular age/sex classes for the cats in three study colonies. More detailed examination of the quality of social behaviour in such colonies confirms that the interactions are highly structured, and the colonies are far more than simple aggregations around food sources.

Tail up

Cats usually precede amicable interactions by raising their tail to the vertical. In feral cats, tail-up precedes both sitting down with another cat and rubbing on another cat (see Fig. 8.3). Pet cats approach a tail-raised silhouette of a cat faster than one with its tail down, confirming that the vertical tail signals an intention to interact amicably (Bradshaw and Cameron-Beaumont, 2000).

Allogrooming

Cats spend a great deal of time grooming, and there is no evidence to suggest that a solitary cat is any less clean than a cat that is groomed by others. The function of allogrooming, which is most often directed to the head and neck area of the recipient (Van den Bos, 1998), is therefore likely to be primarily a social one, except in the case of young kittens that are groomed by their mother before they become competent at grooming themselves. In one study of allogrooming in an indoor colony of neutered adults (Van den Bos, 1998) the flow of allogrooming bouts within pairs of cats was asymmetric, with cats showing more aggression grooming rather than the other way round. Grooming was also associated with aggressive behaviour in 35% of interactions (with groomers showing aggressive behaviour more often than groomees). Other colony studies also indicate that the higher the density of cats the less aggression and the more allogrooming occurs, suggesting that allogrooming may be a way of redirecting potential aggression and reducing tension between cats living together. Although Van den Bos could not detect any effect of degree of relatedness on frequency or duration of grooming, Curtis *et al.* (2003) studying a different neutered colony of cats found kin preference in both allogrooming and proximity.

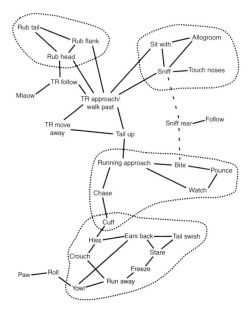

Fig. 8.3. A classification of social behaviour patterns performed by neutered cats (i.e. excluding sexual behaviour). Patterns that are very likely to be performed by the same cat during a single interaction with another cat in its own social group are joined by solid lines. The patterns fall into clusters, most of which have self-evident functions. Amicable interactions (top) often begin with the tail-up signal (centre) either when the cat is stationary (tail up) or moving (TR): these fall into two types, one consisting of allorubbing (top left) and the other mutual sniffing and grooming (top right). At the bottom is a defensive cluster, and above it a cluster of aggressive patterns, linked together by chase/cuff. Of the patterns not strongly linked to any of these groups, miaow may be an attempt by one cat to alert another that it is following with the intention of initiating a bout of rubbing; paw may be an attempt at initiating an interaction which triggers aggression in the other cat and is followed by roll and then defensive behaviour; follow/sniff rear (only weakly linked to other patterns, shown by broken lines) may be followed by amicable sitting together or by aggressive behaviour, presumably depending upon the reaction of the cat being sniffed. (Data collected by Sarah Brown and analysed by Charlotte Cameron-Beaumont, derived from 2044 interactive sequences between 42 neutered cats in three permanent groups. Solid lines represent positive 2 × 2 associations; $P < 0.001$ by chi-square.)

Allorubbing

There is increasing evidence that one of the key behaviour patterns that cements an existing cat group together is mutual rubbing, in which pairs of cats rub their foreheads, cheeks, flanks and sometimes tails together. There may be two communicative aspects to this behaviour, one being the tactile signals exchanged and the other the potential mixing of the two cats' individual scents. So far neither aspect has been investigated separately, so that for the present the function of this display has to be deduced from the behaviour of the animals that initiate it, and the behaviour patterns that precede and follow it in the course of a social interaction.

Several studies have shown that rubbing flows markedly asymmetrically between members of a colony of cats, with females initiating rubbing more than males, and young animals initiating more rubs than older ones (Fig. 8.4). For kittens, rubbing may indicate the strength of their relationship towards the lactating females that suckle them, because a close relationship has been found between the number of times that each kitten rubs on a particular female and the number of nursing bouts it receives from that female. It is unclear which participant initiates interaction – does each female permit a kitten to suckle in proportion to the number of rubs she has received, or does each kitten rub in direct response to being nursed? Whichever way round this is, rubbing does seem to have special significance as far as nursing is concerned; for example, the levels of grooming and nursing that females give to individual kittens in a pooled litter are not closely related.

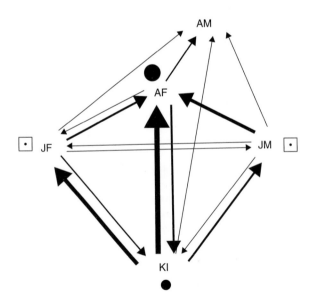

Fig. 8.4. The flow of allorubbing between the age/sex classes of farm cats in a breeding colony at Barleypark Farm, Oxfordshire, UK. The proportions of interactions involving rubbing are indicated by the width of the arrows (between age/sex classes) and the radii of the circles (within age/sex classes). Small circles are shown inside squares, for clarity. The proportions are not corrected for the numbers of individual cats in each age/sex class; average numbers are shown in brackets. AM, adult males (13); AF, adult females (33); JF, juvenile females (9); JM, juvenile males (6); KI, kittens of both sexes (12) (from unpublished data collected by Warner Passanisi and David Macdonald).

The significance of allorubbing between adults can be explored further by examining the behavioural context in which it occurs. Two studies of neutered feral colonies showed that an individual that is about to rub almost always raises its tail as it approaches the other cat (Brown, 1993; Fig. 8.5). The form of the rubbing itself depended upon whether the receiving cat also raised its tail. If it did, both cats usually rubbed simultaneously; if not, the recipient cat sometimes only rubbed after the initiator had, or not at all. Females and males may use rubbing differently: in a neutered colony of (presumably) related cats, Cafazzo and Natoli (2009) found that while tail up and rubbing were generally displayed by females towards males, sniffing nose was more often displayed by males towards females. However, Barry and Crowell-Davis (1999) did not observe any allorubbing in 20 neutered pairs of female indoor cats, possibly because the majority of these pairs consisted of unrelated individuals.

Anthropomorphically, rubbing seems to be highly affectionate but it appears to be used predominantly when the relationship is somewhat one-sided, and may be the nearest the cat has to a behaviour pattern that is used between individuals to reduce conflict through reinforcement of affiliative bonds.

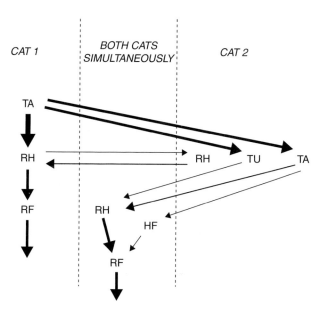

Fig. 8.5. Typical sequences of behaviour that contain head or forehead rubbing between cats. Cat 1 approaches Cat 2 with its tail raised (TA). If Cat 2 does not raise its own tail, Cat 1 rubs its head (RH) on Cat 2, which may reciprocate before Cat 1 rubs its flank (RF) on Cat 2. If Cat 2 does raise its tail (TU) or approaches Cat 1 with its tail raised, they simultaneously rub heads or foreheads (HF) together, before rubbing flanks together. Widths of arrows are proportional to frequencies (except for those emanating from RF). All transitions derived by first-order Markov chain analysis, excluding those with probabilities worse than 0.001 by chi-square (Brown, 1993).

However, it is unlike the appeasement behaviours shown by other social carnivores to avoid escalating aggression during interactions, since rubbing is rarely
seen in interactions involving any kind of overt aggression. Exceptions to this
have been recorded, as when a particular tomcat was occasionally mildly aggressive towards a female that persistently attempted to rub against him (Macdonald
et al., 1987), but even in this case there was no evidence to suggest that the
female rubbed to appease the tom.

Scent marking

Social odours feature prominently in the lives of many mammalian species.
Scents can be specific to a particular individual, are fairly stable with time and
offer the considerable advantage that they can be deposited in the environment
and later detected and decoded by a conspecific in the absence of the emitter.
The scents known to be used as marks by cats include those carried in and by the
urine and faeces, and those originating in skin glands on the head. Urine scent
marks are known to convey individual- and group-specific information (described
in more detail in Chapter 5, this volume), although the way this information is
used in social interactions is uncertain. Male cats frequently spray urine when
consorting with an oestrous female, and it is possible that the rate of spraying
is an indicator of mating success. Any possible role of this scent marking in the
selection of sexual partners either by the male or the female remains unclear.

Social Structure

The communicative repertoire of the domestic cat, although not as complex as
that of the most social carnivores – the wolf, for example – nevertheless suggests
the existence of a social system into which all social interactions should fit. There
is significant disagreement as to how that system should be conceptualized. The
earliest studies of cat sociality, performed on laboratory colonies, used the dominance or 'peck order' concept to derive hierarchies, based on priority of access to
resources such as food (e.g. Winslow, 1938). However, when behavioural ecologists began to study free-living cat colonies, they could find little evidence for
dominance hierarchies, apart from older males preventing younger males from
breeding: cat society appeared to revolve around coalitions between females,
which were usually egalitarian (Macdonald *et al.*, 1987). Both of these approaches
have been used in the interpretation of social interactions within multi-cat households: some authors (e.g. Crowell-Davis *et al.*, 2004; Cafazzo and Natoli, 2009;
Fig 8.6a) have espoused the hierarchy, while others (e.g. Rochlitz, 2005b) have
expressed doubts as to whether it is useful to apply this approach to pet cats.
The usual alternative approach involves focusing on the way that each cat uses
the space available (Bernstein and Strack, 1996; Bradshaw and Lovett, 2003;
Fig. 8.6b), based on the idea that cats are fundamentally territorial animals.

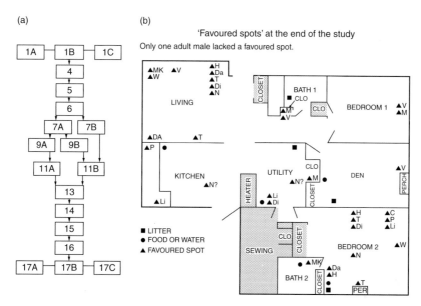

Fig. 8.6. Contrasting approaches to the characterization of social structure in multi-cat households. (a) A dominance hierarchy based on agonistic interactions (from Knowles *et al.*, 2004); each box represents a different cat (*N* = 19 out of 27 cats present) and the numbers indicate dominance rank, suffixed by letters indicating cats of equivalent rank. (b) Favoured resting locations for 13 cats (from Bernstein and Strack, 1996); most cats had several of these and many were shared between two or more individuals.

Dominance can be a loosely defined concept (Drews, 1993) and also one that may mean more to the human observer than to the animals themselves (Appleby, 1993). In any group of animals, it is usually possible to measure consistent asymmetries in the exchange of competitive behaviour between pairs of animals. Combining these dyadic interactions into an overall structure can reveal apparent hierarchies – for example, a more or less linear hierarchy in which one individual is dominant over all others in the group and, at the lower end, one or more are subordinate to all others. Robust mathematical techniques now exist for the construction of such hierarchies (e.g. Bang *et al.*, 2010), but it is often unclear whether these constructs reflect everything that is of importance to the animals themselves, or whether they are an artificial construct created by the human observer. First of all, the 'hierarchy' can change markedly depending on which behaviour pattern(s) are used to construct the dominance relationships (e.g. Natoli *et al.*, 2001), and the context in which interactions are observed. Secondly, the fact that a hierarchy may be apparent to the human observer does not mean that it is also apparent to the cats themselves, or that position in the hierarchy is something to which each cat aspires. In other words, some groups of cats may behave *as if* they inhabit a hierarchy, but are in fact simply reacting to the cats around them in their own individual way, or as a consequence of learning from previous encounters the optimal manner in which to interact with others. However, the cats themselves are unlikely to be aiming to achieve a species-specific structure, adopting any particular 'role' within that structure, or determining behaviours based on changing their role within the structure. Indeed, because today's domestic cats are only a few thousand generations removed from their solitary territorial ancestor, and hunting remains a solitary activity, it is difficult to see how a species-specific social structure could have evolved, nor its value in determining social behaviour.

The apparent hierarchies that can be recorded from cats kept in high-density colonies may simply reflect stress brought on by overcrowding: it has been suggested that the 'lowest-ranking' cats in these groups would normally disperse (Durr and Smith, 1997). In feral breeding colonies, cooperation between females has been described as 'centripetal' and not 'hierarchical' (Macdonald *et al.*, 1987), but functional hierarchies can often be derived among males, in which older males harass younger males, especially juveniles, thereby preventing them from mating. Adult male cats rarely, if ever, cooperate, and this hierarchy appears to be a simple 'pecking order' in which younger males learn to avoid older, more experienced males with which they cannot (yet) effectively compete. Although 'dominant' resident males achieve more matings than 'subordinate' resident males, doubt has been cast on whether older males actually have more reproductive success (Say and Pontier, 2004): females show strong preferences for non-kin males (Ishida *et al.*, 2001), resulting in many – even a majority of – kittens being fathered by males from outside the colony (Yamane, 1998). Thus the most successful strategy for a male cat would appear to be to visit any neighbouring colonies as often as possible. On such visits, even males that appear dominant in their own colony behave like subordinates, suggesting that 'dominance' is not an individual characteristic but a consequence of the learnt interactions within familiar groups. In multi-cat households, even neutered cats can behave in a 'bullying' way, thus affecting the movement or activity of the other cats (see Chapter 11, this volume). However, the behaviour of the major-ity of cats in such agglomerations appears to be regulated by mutual avoidance, time-sharing and, for some pairs of (especially related) individuals, active affilia-tion (Bernstein and Strack, 1996), possibly derived from normal female–female behaviour.

Most of the behaviour patterns that have been put forward as indicators of dominance and submission are part of normal offensive and defensive behav-iour, as occurs in other contexts, for example between unfamiliar individuals. In some but by no means all colonies, older males mount younger males, but it is not clear whether this is dominance-assertion or merely an outlet for sexual tension in the proximity of unreceptive females (Yamane, 1999). Feldman (1994b) has reported juvenile males in one colony performing the rolling behaviour normally characteristic of oestrous females, apparently as a submissive behaviour towards older males, but this does not seem to occur universally. Macdonald *et al.* (1987) speculated that rubbing behaviour might be an indicator of subordinate status in large colonies, but this has yet to be confirmed by subsequent observations. Overall, it appears that the ritualized (i.e. evolved, species-typical) signals described in other species as determining dominance relationships are missing from the domestic cat's repertoire. This is consistent with the idea that they have no species-typical social structure other than whatever emerges as a consequence of each cat learning about the likely responses of others under different circumstances. In this conception, Macdonald *et al.*'s 'centripetal' system can be thought of as arising from the persistence into adulthood of normal affiliative relationships between mothers and their female offspring.

The Functions of Domestic Cat Sociality

There has been a good deal of speculation over the advantages that cats might gain from living in groups, but to date not a great deal of evidence has been found for any of the alternatives. Given the wide range of group sizes and the highly artificial surroundings in which some of the largest groups find themselves, it is probably not surprising that a functional explanation that appears to apply to a group in one type of situation does not apply in another. Certainly a great deal of care is needed when extrapolating from the sociality of domestic cats to that of wild felids. Social structures in wild species of Carnivora have been refined by millions of years of natural selection; not only is *Felis catus* only a few thousand generations old, but the circumstances under which it has evolved have not been consistent, as the niches supplied by man's activities have continually changed.

To recap, cooperation appears to revolve around the activities of females attempting to pool their ability to defend resources, and thereby obtain maximum benefit for their offspring from the concentrated food source that has allowed the colony to become established. When all the females within a group are closely related, and they usually are, this can be extended to include their relatives' offspring. The best evidence for this comes from the much poorer breeding success of peripheral females compared with central females at Horspath Farm (Kerby and Macdonald, 1988). At that farm this crucial spatial status was determined by kinship. The two central lineages had much higher breeding success than the peripheral lineages (even though some central individuals would occasionally breed in peripheral sites), which consisted of four side-branches of the original central lineages and two lineages based on immigrants. After several generations, these peripheral lineages would probably die out, or emigrate; meanwhile, if the central lineage bred successfully, it would tend to fragment and push some of its females to peripheral positions. In practice, this process is likely to be disturbed by man's activities (culling, the taking of kittens for pets, changes in the amount of food and shelter available), and so is unlikely to be readily apparent in all colonies.

The Origins of Sociality in the Domestic Cat

Since there are no published accounts of the behaviour of group-living *F. silvestris* apart from the domestic cat, we can only speculate as to how the sociality of which this species is evidently capable arose in the first place. Virtually all the social groups that have been studied have relied on concentrations of food supplied by man. This raises the possibility that sociality in the domestic cat has arisen secondarily, as a by-product of domestication; *F. s. lybica* is essentially solitary and territorial. If we assume that *F. s. lybica* originally exploited anthropogenic environments as a commensal, and that those niches contained reliable sources of prey that could feed more than one queen and litter, then individual animals capable of forming coalitions with kin might out-compete individuals that could not. This would pave the way for the rapid evolution of cooperative behaviour. Subsequently, as the cat was domesticated, man might have deliberately chosen individuals that tolerated the close proximity of other cats, because one highly territorial cat would not have achieved the desired effect. Later on, those cats that also displayed affiliative behaviour towards people might have been selected from the original, conspecific-tolerant, population. The affiliative behaviours that adult cats direct towards each other and towards people may have been derived from those shown by kittens towards their mothers, carried into the adult state by a process of progressive artificial neotenization.

The Cat–Human Relationship　　9

Introduction

The cat–human relationship has features that set it apart from any other type of human–animal interaction. The closest analogy that can be made is to the dog–human relationship, in which the dog appears to use various elements of the social behaviour of its wild ancestor, the wolf, to communicate with man (Miklósi, 2007). The idea that will underlie the whole of this chapter is that the cat also treats its human owner as if it were a member of its social group. Unfortunately, because the social systems of the various races of *Felis silvestris* are much less well understood than those of the wolf and dog, comparisons between cat–cat and cat–human behaviour must necessarily be speculative. However, some general similarities are apparent. For example, the absence of any consistent patterns of interaction between cats within social groups may have its counterpart in the 'independent' relationship that cats have with their owners.

Where cats and people live together in the domestic environment, the relationship appears to have an impact on the humans as well as the cats. The impact of cat-keeping on physical and psychological well-being in people has been the subject of some research, but is outside the scope of this text. More information is available from reviews in Turner and Bateson (2000), Bernstein (2007) and O'Haire (2010).

The impact of living with people on cat social and territorial behaviour

The degree of sociality shown by cats towards conspecifics is profoundly affected by food distribution, free access to space and population density (see Chapter 8,

© J.W.S. Bradshaw, R.A. Casey and S.L. Brown 2012. *The Behaviour of the Domestic Cat*, 2nd Edn (J.W.S. Bradshaw *et al.*)

this volume). These factors impinge very differently on a pet cat as compared with a semi-wild feral or farm cat. Most cats living with people benefit from a predictable source of food, which impacts on hunting behaviour (see Chapter 7, this volume), and may influence the desire to range. However, the amount of space which owned domestic cats can occupy varies considerably with circumstance. A pet cat living in the countryside may enjoy similar freedom of space to an independent farm cat. In contrast, many town-dwelling cats have their freedom of movement restricted, either because they are kept as indoor-only pets, because their owners control their access to the outdoors or, less obviously, because their free movement outside is inhibited by the local population density of cats. The implications of this wide variation in the size of home ranges on behaviour will be discussed in Chapter 11. Although there does not seem to be any evidence that a cat that has always been restricted to living in an apartment must necessarily behave abnormally, some authors suggest a higher incidence of behavioural 'problems' reported by the owners of indoor cats. Potentially of greater impact is the restriction of owned cats in their choice of cat companions. Cats living in semi-independent social groups are generally able to choose with which other cats they interact; pet cats rarely have this degree of choice, even in multi-cat households, where the owners will usually select which cats will live together. As discussed further in Chapter 11, living in close proximity with other individuals that are not socially compatible is a common factor in the development of undesired behaviour, and a welfare concern. The option to leave the household and try to establish a home base elsewhere is as available to pet cats as it is to ferals, and there is much apocryphal evidence that this does indeed happen, although little scientific study seems to have been carried out. In addition to the problems of forced contact with incompatible cats, it is unknown to what extent absence of affiliative interaction with their own kind impacts on normal behaviour and welfare for domestic cats that are housed singly. There is no indication that cats suffer from being denied the company of members of their own species, although their situation is a highly artificial one. This is because the solitary lifestyle, which would normally only occur at very low population densities, is in this case often combined with a high density of more-or-less inimical cats. The possible effects of this on territorial behaviour will be discussed below.

Cat–Human Behaviour

Cats do not generally follow their owners about in the way that dogs do, and so interactions with their owners tend to consist of short bouts. Dennis Turner (1991) has shown that interactions that are initiated by the cat tend to be longer than those initiated by the owner, particularly when the owner is very active in starting the interactions. His data suggest that people who are over-keen to extract as much as possible from their relationship with a cat run the risk of actually spending less time with that cat than they would in a more casual relationship in which the cat is allowed to make much of the running. Indeed, the common suggestion that cats often attempt to interact with people who profess a dislike of them may reflect their preference for those who do not seek to interact. This was supported in a study showing that owner personality traits can affect temporal patterning of human and cat behaviour and its complexity (Wedl *et al.*, 2011): owners with high scores for the trait 'neuroticism' had less frequent interactions with their cats. Such owners appear to be likely to seek more contact with their cats (i.e. take the initiative in interacting with their cats), possibly resulting in their cats being correspondingly less active contact seekers. Cats' preferred patterns of interaction with people seem analogous to those with other cats, where interactions may be relatively frequent but of short duration and intensity. This can be at odds with the type of interaction preferred by cat lovers, who may be out for prolonged periods and then seek high-intensity interactions when they are at home.

Cats often prefer to interact with one particular member of the household in which they live. In families with children, Claudia Mertens (1991) has found that while children tend to approach the family cat more than their parents do, the cat itself prefers to interact with the adults. Moreover, in general, the cat–human relationship is more intense when the human partner is female. The most obvious explanation for this is that it is often a female member of the household that feeds the cat. While cats seem to know which member of the family is most likely to provide food, and direct their interactions towards that person when they are hungry, they are likely to be just as affectionate towards other family members at other times. For example, research by Geering (see Turner, 2000) suggests that the act of feeding a cat can enhance the establishment of a relationship but is not sufficient to maintain it. Other interactions (stroking, playing, vocalizing, etc.) may be required to cement a newly founded relationship.

Cat–human communication

Displays of behaviour performed by cats towards their owners contain a variety of visual, vocal, tactile and olfactory signals. Bradshaw and Cook (1996), observing cats prior to being fed, reported that some of their behaviours were evidently directed at the owner and some at objects, while others were not obviously orientated. All the behaviour patterns they recorded were also observed in adult cat–cat interactions, suggesting they are derived from normal social signals of group-living cats.

It is self-evident that a cat wishing to defend itself against a person will hiss and strike out as it would towards an attacking cat or, for that matter, a dog. It is probably not so obvious that a cat rubbing around its owner's legs is behaving like a juvenile cat towards an adult female (see Chapter 8, this volume). Some of the human-directed patterns, such as purring and kneading with the paws, have been classified as infantile; while the latter is probably derived from the treading action with which suckling kittens stimulate the flow of milk from their mother, purring is now known to occur in a wide range of contexts even among adult cats (see Chapter 5, this volume) and must be regarded as a pattern occurring throughout life.

Cats use a wide range of other vocalizations towards humans, and some owners claim to be able to recognize a wide range of meaning and nuance in these calls – this is discussed in more detail in Chapter 5. Little is known about the significance of any of these calls in cat society, and hence it is difficult to put cat–human vocal communication in an ethological context. However, cats' much greater use of vocalizations towards humans than towards other cats strongly suggests that cats quickly learn that humans respond positively towards mid- to high-pitched sounds like the miaow and the 'solicitation purr' (McComb *et al.*, 2009).

The vocalizations used by humans when interacting with cats are often similar to those directed at infants, sometimes referred to as 'motherese'. Sims and Chin (2002) looked at vocalizations used by humans towards cats when engaging an unfamiliar cat to play with a toy. The students involved almost all spoke to the cat, using language similar to that used towards children (i.e. short words, repetition and 'imperatives'). They concluded that humans in this situation modify their speech based on their perceived comprehension of the listener.

Allorubbing, which appears to be both an amicable and a mildly asymmetric action (see Chapter 8, this volume), is commonly directed by cats towards their owners, and it is possible that stroking is the nearest human equivalent to the reciprocation often seen in the cat–cat context. Cats have a preference as to exactly where on their body they like to be stroked by their owner (Fig. 9.1), their favourite place being the temporal gland, the cheek area between the eye and the ear, and their least favourite being in the caudal gland region around the tail. A survey by Bernstein (2007) found that 48% of cats prefer petting in the head area and only 8% prefer to be petted on the stomach or tail. Cats appear to indicate which areas they would like to be petted in by engaging in behaviours such as staying still, closing their eyes or positioning their body so as to encourage rubbing of specific sites. Owners describe cats using behaviour patterns that can be interpreted as seeking initiation of petting (e.g. jumping on lap, rubbing on person's leg). These patterns of interaction are indicative of some form of individualistic ritual (Bernstein, 2007), for which learning is likely to play a major role.

Interestingly, cats that are allowed out of doors rub on their owners more frequently than cats confined indoors, and cats in multi-cat households tend to rub less on their owners than do single cats (Mertens, 1991); this may be because rubbing is often used as a greeting behaviour pattern, after a cat has been absent from its social group for some time. Tame specimens of *Felis lybica* also allorub their owners (Smithers, 1968), even though they are in general much less domesticated than *Felis catus*, so this behaviour pattern was almost certainly part of the behavioural repertoire of the earliest domestic cats.

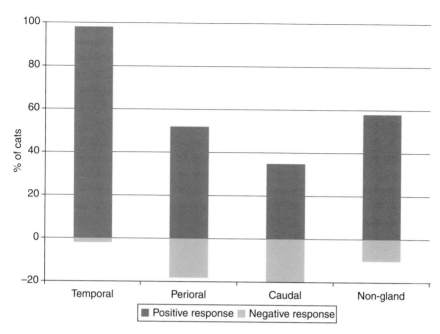

Fig. 9.1. Positive and negative responses (%) of cats to stimulation of temporal (upper cheek/temple, between eye and ear), perioral (chin and lips), caudal (on lower back at base of tail) and non-gland areas (data from Soennichsen and Chamove, 2002).

Claudia Mertens and Dennis Turner (1988) demonstrated that the use of vocalizations and rubbing by the cat changes depending on whether the human responds. They allowed a cat to enter a room that contained an unfamiliar person; for the first 5 min that person was not allowed to return the cat's attempts to interact, while for the next 5 min he or she could interact without any restrictions on the form of that interaction. The level of vocalization halved in the second phase compared with the first, suggesting that the cat had initially been calling to try to induce the person to interact. In the second phase the rate of head rubbing on the person increased fourfold, while the number of flank rubs increased only slightly. However, in similar trials, but with different cats and a familiar person (herself), one of us (SLB, unpublished) found a similar increase in head rubs when she began to interact, but mainly directed at nearby objects rather than at her.

Although they do not generally avoid eye contact with humans, cats have been shown to use eye contact less frequently than dogs during interactions with people. Miklósi *et al.* (2005) suggest that the lack of frequent and early glances at humans shown in their studies of cats may be a sign of relative independence from humans compared with dogs, and may also contribute to differences in trainability between domestic cats and dogs.

Territorial behaviour

Theoretically, since maintenance of a territory has the function of enabling access to sufficient space for hunting, pet cats should have less of a need to defend a territory than ferals, because they have a reliable source of food provided by their owners. However, it is self-evident that many cats actively defend a territory centred around their owner's house from other cats in the neighbourhood. This does not prevent cats from entering each other's houses through the cat-flaps or windows and stealing one another's food when the opportunity arises. Indeed, cats' apparent willingness and ability to defend territory varies considerably between individuals. Although some cats appear to chase other cats away from the area around their owners' houses, other cats' territories may not extend any further than the interior of their owners' houses. When venturing outdoors, such cats may take long and circuitous routes in the spaces between the territories of other cats to reach areas, such as patches of urban woodland, that contain no residents (Fig. 9.2). Others may simply stay

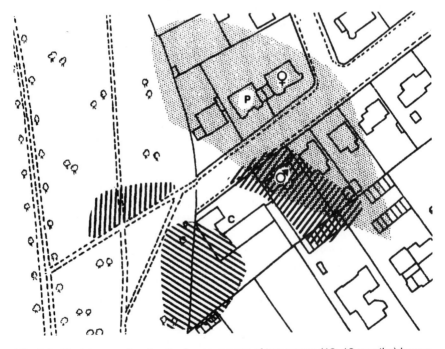

Fig. 9.2. Sketch map showing the home ranges of two young (12–18 months) house cats, one a neutered male (striped areas) and the other a neutered female (stippled area), obtained by radio-tracking. The houses that these cats lived in are denoted by the male and female symbols. Two other cats, both neutered males, lived in adjacent houses (C and P): both latter cats were observed attacking the young male. This male was never sighted in transit between its three sub-ranges, so these have been kept separate to avoid giving a false impression of the total area that it used.

close to home, ready to avoid confrontations by immediately retreating into their owners' houses.

Where studies have compared the movements and use of space by owned and unowned cats simultaneously, pet cats appear to have smaller home ranges than unowned cats (Schmidt *et al.*, 2007; Horn *et al.*, 2011; see Table 9.1). From these studies it appears that home ranges decrease in size as provision of food increases. Kays and DeWan (2004) found an average home range of just 0.24 ha for house cats in their study; these ranges overlapped considerably, with each cat ranging over an average of 3.67 gardens including that of their owners. Ranges also vary according to where the cat and owner live. Lilith *et al.* (2008) found a range of 0.07–2.86 ha for pet cats from rural areas and 0.01–0.64 ha for urban cats.

Many studies have noted a high inter-individual variation in home range size of house cats that seems to be unrelated to gender or hunting success (Barratt, 1997; Meek, 2003; Metsers *et al.*, 2010). Meek classifies this into two different types, 'wandering' (those that roamed outside of their household domain (>1 ha)) and 'sedentary' cats. The reason why some roam and some do not is unclear – some cats have a preference for certain prey types that may not always be close to their home, which may explain a larger home range for some. Possibly it is simply another aspect of the different behavioural styles of different individuals.

Although male ranges tend to be larger than female ranges in sexually entire cats, little difference between the two has been noted in most studies of pet house cats. This may be explained by the neutered status of the majority of pet cats, for whom the reproductive strategies that affect space use in entire cats (see Chapter 8, this volume) are no longer a factor.

Home ranges of cats living in the same house appear to overlap considerably (Barratt, 1997; Meek, 2003), and more so if they are related (Barratt, 1997). Evidence for overlap in home ranges of males from separate residences and males and females from separate residences has also been found – however, these cats may actively avoid each other through spatial detail and timing of their movements (Barratt, 1997; Meek, 2003). Barratt (1997) noted, however, that there was no overlap of females from separate residences, possibly implying some kind of group territory.

The importance of territories on the activity and behaviour of cats in the domestic environment is highlighted when there are changes in cat populations

Table 9.1. Home ranges (ha) calculated for two studies that monitored movements of owned and unowned cats simultaneously in their study regions.

Research group	Owned cats	Semi-feral cats (unowned but fed)	Unowned feral cats
Schmidt *et al.* (2007)	1.6	5.3	14.7
Horn *et al.* (2011)	1.9	n/a	94.2

within an area. The introduction of new cats into a neighbourhood can have consequences for cats across a wide area, as the activity of each cat is influenced by changes in its neighbours' patterns of activity. Since established patterns of spatial and temporal avoidance may be disrupted, these changes may result in bouts of overt aggression where unexpected contact occurs. Changes may also result in the reduction of ranging opportunities for some cats, potentially impacting on their behaviour inside the house (see Chapter 11, this volume). With permanent or temporary changes in cat populations occurring regularly, for example with cats going in and out of catteries or veterinary practices, such fluctuations of territory and relative contact with other cats are likely to be a common source of anxiety and behavioural change for the domestic cat.

The Socialization Process

What is known about the process whereby cats become socialized to their conspecifics has been described in Chapter 4. Cats, like dogs, readily become socialized to more than one species; this capacity seems to be unique to the domesticated Carnivora, since other companion animals, such as equids and rodents, do not show the same degree of attachment to humans. Kittens of the right age will socialize towards almost any mammal with which they come into close and amicable contact, although there is some evidence that bonding to other kittens is slightly stronger than bonding to other species. For example, if a puppy is raised with a group of kittens, and each kitten is separated from its littermates so that it cries to re-establish contact, the crying is more reliably suppressed by the arrival of another of the kittens than by the puppy (Kuo, 1960). This does not seem to be simply due to the way that the puppy behaves in that situation, because a kitten raised with five puppies finds each one of those puppies an effective comforter. In fact, the suppression of crying in these kittens (which know only puppies) is equally as effective as the suppression of crying by a kitten in artificial litters that contain only one puppy.

The socialization of cats towards humans has been the subject of a number of studies, described below, which collectively show that the process is not all or nothing, but contains variable elements. For example, the amount of handling a cat receives, the age at which it occurs and the number of handlers all affect the degree of friendliness towards people later in life.

The sensitive period

Socialization is most effective when the kitten is young, and so is at least superficially similar to the imprinting process in precocial birds, such as ducklings. Kittens develop much more slowly than ducklings, and so it is not surprising that they are capable of forming social attachments over an extended period of time. Because this is a much less constrained process than imprinting, the term 'sensitive phase' is often used to describe it. Within this phase attachments are formed to any object with the right stimulus qualities (which have not been particularly well defined for the cat, but must include complex form and texture, and movement). One sensitive period occurs quite naturally in the young kitten, but even in older cats some attachments can form, or preferences for species partners change. One striking example of this is the attachment that can be formed during severe illness; there are many anecdotal accounts of previously wild feral cats forming strong bonds with people who have nursed them through recovery from injury or disease. This reactivation of the socialization process is apparently induced by intense stress.

The precise timing of the 'normal' sensitive period in kittens was elucidated through a series of experiments where batches of kittens were handled for periods of 4 weeks, with different groups starting in different weeks. From this, Karsh and Turner (1988) identified the sensitive period for socialization in cats as beginning at 2 weeks of age and ending at about 7 weeks. If socialization has begun during this sensitive period a cat's reactions to humans will continue to change over the next couple of months. Lowe and Bradshaw (2002) found that in pet cats the amount of handling received during the sensitive period produced a differential effect on their reaction to handling at 4 months old but not thereafter, suggesting that this aspect of the cat's relationship with people continues to take shape throughout the first 4 months of its life. A variety of experience during the whole socialization period should therefore affect the quality of the cat's subsequent relationship with humans.

Up to a certain limit, the more handling a kitten receives the friendlier it will be towards humans (Turner, 2000); Karsh (see Turner, 2000) found that kittens handled for 40 min/day subsequently approached a person more quickly and could be held for longer than kittens handled for 15 min/day, and positive effects of handling of kittens in animal shelters are detectable in the cat's temperament at 1 year of age (Casey and Bradshaw, 2008). However, more than 1 h/day of handling produces no further significant increment in friendliness (see McCune *et al.*, 1995; Turner, 2000). Karsh did find, however, that her results were greatly affected by the 'personality' of the kitten, and that the effect of the sensitive period on the change in sociability was actually rather greater in the most timid kittens than in the others. Personality is examined in more detail below.

The number of different handlers that a kitten experiences can also determine its reaction to strangers. Cats that have been handled by only one person can be held for, on average, twice as long by that person than by any other, but cats with experience of four handlers will stay with any person, including a stranger, for the same length of time as the 'one-person' cats will stay with their handler. Thus there seems to be an element of generalization after several humans have been encountered, such that the one-person cat is socialized to that individual, while the multi-person cat becomes socialized to all humans that behave in broadly the same way. However, Collard (1967) found that while kittens handled by five people were more outgoing, one-person kittens were more affectionate towards their handlers, purring more often and play-ing for longer periods. These differences may reflect the different pathways of development that a kitten can go through, depending on the availability and character of adult conspecifics, which may influence its later social (cat–cat) interactions. However, as was discussed at the end of the previous chapter, very little is known about these processes in any context. It has also been suggested that experience of specific stimuli, such as contact with handlers of a particular sex, will differentially affect perception of men and women when kittens reach adulthood (McCune *et al.*, 1995).

The influence of the mother cat in the socialization process should not be underestimated. All of the studies described above were done under highly con-trolled conditions, with maternal influences eliminated or kept to a minimum. It sometimes happens that queens give birth in an inaccessible place and then discourage human (and indeed cat) access to the kittens until they are several weeks old. By doing this, she may well place limits on the degree to which they can be socialized to man. Her more indirect influences, which will almost cer-tainly include the kittens imitating her reaction to people, have received little study, but are probably an important factor determining each kitten's future relationships with the human race.

Cat 'Personality'

Every cat is an individual. This is self-evident to owners and is one of the rea-sons the cat is so fascinating as a pet.

The consequences of individual differences, in terms of their contribution to a lifetime's reproductive success, are extremely difficult to evaluate, and have been speculated about much more than studied directly. Mendl and Harcourt (1988) have suggested three general reasons why individual differences should persist in a population. One is that two strategies might simply be equally suc-cessful, and therefore neither would confer a disadvantage. Another possibility is that each strategy is the most successful under a different set of circumstances, and if those circumstances are unpredictable in either when or where they may occur, all the strategies will tend to persist in some individuals in the whole pop-ulation. Thirdly, it may pay individuals of different size to behave differently; larger animals may be able to maximize their cost/benefit strategies, whereas smaller individuals may have to 'make the best of a bad job'. The various mat-ing strategies of feral male cats may come into the latter category. However, the domestic cat may not be an ideal model for exploring the consequences of individual variation in behaviour; just as domestication has resulted in a great diversity of coat colours, so it may also have led to greater diversity of 'person-alities' than in a wild species.

Whatever the evolutionary and developmental causes of individual varia-tions in behaviour, they presumably affect all aspects of a cat's life; for example, one aspect of the feline personality is the degree of equability towards other cats.

However, it is the effect of personality on the human–cat relationship that has received the most attention, which is why this topic is being dealt with in this chapter, rather than another.

It has been shown that human observers can reliably rate cat personalities, which correspond to objective measures of the relative frequencies of various behaviour patterns that each individual expresses under identical circumstances (Feaver *et al.*, 1986).

Simple distinctions between personalities have been made from cats' reactions to just a single set of circumstances. For example, Mertens and Turner (1988) divided cats into three categories based upon their approach to unfamiliar persons: initiative/friendly, reserved/friendly (depending on whether the cat or the human was the more likely to start the interaction) and unfriendly types. Turner (1991) distinguished two types of friendly personality among Swiss house cats, one preferring play contact and the other petting.

More detailed research suggests that adult cats can be separated into three main types, classified in terms of their individuality with respect to their interactions with people (summarized in Table 9.2). Type 1 personalities tend to be those cats that have a very positive relationship with their owner, and also tend to approach unfamiliar people. A second type can be timid, shy, unfriendly and nervous. The third type of cat is active and may also tend to be aggressive.

Other than its individual style or personality, other aspects of the cat itself such as age or sex do not seem to influence the relationship between cat and human. Adamelli *et al.* (2005) did find that a cat's behaviour towards its owner was influenced by the number of people in the family, by the presence of children and the people who looked after the cat – the most sociable cats in this study were those living in small families without children, a trend also detected by Mertens (1991), who found that interaction, proximity and rubbing by the cat were moderately more frequent in smaller than in larger families. Turner (2000) discusses in more detail the various influences concerning the cat, the owner and the housing conditions that have been analysed with respect to their effects on cat behaviour.

Table 9.2. Summary of different cat personality types as described by different research groups.

Type 1	Type 2	Type 3	Researcher(s)
'Sociable, confident, easy-going'	'Timid, nervous'	'Aggressive'	Feaver *et al.* (1986)
'Confident'	'Timid'	'Active'	Karsh (1983)
'Trusting'	'Shy and unfriendly'		Meier and Turner (1985)
'Initiating friendly interactions'	'Shy and unfriendly'		Mertens and Turner (1988)
'Bold'			McCune (1992, 1995)

The origins of 'personality'

Many of the differences that we can perceive between individual cats probably start during the development of the kitten, although as has been emphasized in Chapter 4, this volume, the behavioural development of the cat is a much more flexible and goal-oriented process than was once thought.

While developmental processes are undoubtedly important in determining how friendly a cat will be towards people, it has also been possible to demonstrate both maternal and paternal influences on friendliness (i.e. a friendly mother and father will tend to have friendly kittens). The maternal effect is probably due to a mixture of inherited and direct behavioural effects, but the paternal effect is likely to be entirely inherited, since male cats play little or no part in the rearing of their offspring. Reisner *et al.* (1994) found, in litters sired by five different fathers, that a kitten's paternal genetics affected the time it spent close to a test person.

McCune (1995) investigated the relative contributions of paternity and early socialization on the development of cats' behaviour to people and novel objects. As expected, cats socialized as kittens were friendlier towards a test person than cats not socialized as kittens. Genetic variation also produced significant differences between cats in their response to a familiar person; cats fathered by a 'friendly' father were friendlier to the person than cats from an 'unfriendly' father. However, when it came to approaching a novel object, differences in socialization had no effect while cats from a friendly father were quicker to approach, touch, explore and remain in close contact with the object than cats from an unfriendly father. McCune redefined the genetic contribution to friendliness towards people as 'boldness' – a general response to unfamiliar or novel objects irrespective of whether or not the objects are people. The socialization effect was specific to the cats' response to people.

The relative contribution of genetics and experience to the behavioural expression of the boldness component may be complex – for example, genetically bold kittens could initiate more interactions with people and therefore receive more handling; or kittens could become bold because they experience more handling (or both).

There is some evidence to suggest that, once cat personalities are formed, some elements stay fairly constant for at least several years, possibly for life. Lowe and Bradshaw (2001), studying post-prandial behaviour of cats from nine different litters, found four elements of behavioural style that were consistent for all or part of the period from 4 months to 2 years of age. The traits they labelled 'investigativeness' and 'boldness' were the most consistent, suggesting that they may be stable elements of personality in the cat. 'Staying indoors' (after eating) and 'rubbing' behaviours were somewhat less stable. The elements appeared to develop at different rates, implying more than one underlying process, possibly each a combination of genetics, developmental processes and environmental influences.

The effects of genetic differences on behaviour and subsequent cat–human relationships can be illustrated further by Turner's (2000) work on breed differences in domestic cats. His studies confirmed, through a combination of owner questionnaires and direct observations, that there are differences in cat–owner interactions when comparing pedigree (Siamese and Persian) with non-pedigree cats. Observations showed that purebred cats and their owners spent more time near each other and interacted for longer than did non-purebred 'house cats' and their owners. When asked about their cats, owners rated both of the pure-bred varieties as being more vocal, affectionate towards their owner and friendly towards strangers than the house cats. There were fewer differences between the two pure breeds. More recently Weinstein and Alexander (2010) asked owners of Siamese and non-pedigree cats to rate their cats and also themselves from a list of personality traits. As with Turner's findings, the purebred cats were rated differently from mixed-breed ones. This study also revealed a positive correlation between owners' ratings of themselves and their cats for some of the behaviour traits – for Siamese owners this was clever, emotional and friendly, and for non-pedigree cat owners this was aggressive and emotional traits. Whether owners see themselves as they see their cats, or perhaps choose pets similar to themselves, remains to be elucidated.

The Effects of Neutering

There is an increasing trend for house cats to be neutered at an early age. The behaviour of entire tomcats is reasonably regarded as undesirable by the typical urban cat owner, and many owners are also unwilling to go through the trouble and expense of assisting a queen to raise kittens, as shown each year by the numbers of pregnant queens that are abandoned at cat shelters or simply at a location far enough away from home to prevent their returning. In addition, in order to address the problem of ever-increasing numbers of unwanted pets, many rescue organizations now insist that new owners agree to neuter their cat (if this has not already been carried out) on adoption of the animal from the shelter.

Other than the inhibition of sexual behaviour in toms that have been neutered before the onset of puberty, one of the most notable effects of neutering in male cats is the reduction in aggressive behaviour. The extent to which male behaviour is changed seems to depend to a large extent on the timing of castration. If this is done within the first year of life, urine spraying, aggression towards other cats and roaming are all partially or even totally suppressed; for example, Hart and Barrett (1973) reported that 87% of males reduced their frequency of spraying dramatically after the operation. Roaming may persist in toms neutered at more than 1 year old. Even after a male has had aggressive and sexual experience, castration can reduce aggression, eliminate spraying and increase the frequency of affiliative behaviour patterns such as rolling (de Boer, 1977a). Neither spraying nor fighting is likely to be completely suppressed, but they may take place in locations away from the home base and therefore often go unnoticed. However, in multi-cat households it is sometimes reported that one cat will spray, even if all are neutered, so there appears to be an interaction with the density of cats.

With females, sexual calling is inhibited by the removal of oestrus. The effects of neutering on aggressive behaviour in females are a little less clear, with some studies concluding that neutering has no effect and others suggesting decreased aggression in neutered compared with unneutered females (see Finkler *et al.*, 2011).

Effects on social behaviour

The practice of controlling feral cat colonies by neutering is gaining in popularity (see following chapter). Studies looking at the effects of this on cat social behaviour have generally found a decrease in agonistic interactions, either observed following neutering of a free-roaming group of cats or by comparison with interactions in unneutered groups. Finkler *et al.* (2011) compared behavioural differences between urban feeding groups of neutered and sexually intact free-roaming cats following a trap–neuter–return (TNR) procedure. They found less aggression in the neutered groups and, in particular, fewer agonistic encounters between neutered males. In a concurrent study on the same groups of cats, Gunther *et al.* (2011) found there was more immigration of new cats into the neutered groups. It has been observed in other studies that with neutered groups aggression towards members of neighbouring groups also seems to be reduced. This may point to a risk that, without human assistance, the neutered individuals in a group might be easily displaced by entire cats.

Amicable interactions within neutered groups of cats appear to either remain the same or increase in frequency compared with breeding groups (Brown and Bradshaw, 1996). For example, young males often do not disperse as they would if they had been allowed to reach sexual maturity. However, it is possible that while neutered groups remain cohesive, they are more easily displaced from their core area by immigrant entire cats, which may limit the effectiveness of TNR programmes (see Chapter 10, this volume).

Studying interactions between pairs of neutered indoor cats living together, Barry and Crowell-Davis (1999) found there were no significant differences in affiliative or aggressive behaviour based on the sex of the cat. Neutering is of course only one aspect of the formation of relationships between these cats, with individual differences playing a significant role.

Cat Welfare

10

Introduction

The welfare of animals excites a great deal of public concern, and yet relatively little research has been directed towards the welfare of the domestic cat compared with other domesticated species. However, substantial progress has been made in the past two decades; for a comprehensive discussion, see Rochlitz (2005a). The popular image of the domestic cat is that of a pampered pet, whose every whim is satisfied by its devoted owner. However, reality is often at odds with this image. First of all, considerable numbers of cats live outside the domestic environment for at least part of their lives. This disturbing truth was first brought to light in statistics published by Rowan and Williams (1987), showing that in the USA in the 1980s only one-third of cats remained in the same household for the whole of their adult life, and each year one-quarter of the adult cats left their households; there is little to suggest that the situation is not much the same today. Some cats presumably move by mutual consent of the new and previous owners, but many must either migrate into the stray/feral population, or 'adopt' a new household because of lack of care in their previous home or move because of the local environment with respect to other cats. Add to these the thousands of cats held by rescue organizations (Clark *et al.*, 2012), as well as the millions of cats that lead a shadowy feral existence on the fringes of human society, and the need for more detailed examination of welfare issues becomes self-evident. Secondly, even where cats live in homes, misunderstanding about their behavioural needs and responses can commonly result in compromised welfare, as discussed further in Chapter 11.

© J.W.S. Bradshaw, R.A. Casey and S.L. Brown 2012. *The Behaviour of the Domestic Cat*, 2nd Edn (J.W.S. Bradshaw *et al.*)

How to Define 'Welfare'

Animal welfare can be considered from a variety of standpoints, ethical and philosophical as well as scientific. There is also an important cultural and personal element that influences the degree to which animals are exploited and the degree of animal suffering that is tolerated; for example, in some Far Eastern cultures it is accepted that cats can be killed for their fur or for food (Podberscek, 2009), and accordingly such cats are treated as production animals rather than as companions (PETA, 2011). However, while the human toleration of poor welfare is a subjective issue, considerable progress has been made in establishing a scientific basis for the measurement of welfare, and behavioural considerations are important in both establishing the causes and measuring the effects of poor welfare.

There are three distinct approaches to the application of biological science to animal welfare. The first prioritizes the health and biological functioning of the animal; the second attempts to assess the emotional ('affective') state of the animal, minimizing aversive states such as fear and pain, and (to a lesser extent) maximizing positive states; the third emphasizes the performance of 'natural' behaviour (Fraser, 2009). Since both overt behaviour and internal emotional states have ultimately evolved to improve biological functioning and especially reproductive success, we would expect strong correlations between the three approaches. Although the third can be difficult to apply to highly domesticated animals, since their behaviour may not be adapted to their current environment (deficits, at least in functioning, being compensated for by man), it is arguable that, for the relatively undomesticated cat, all three approaches are valid.

In its simplest form, welfare can be defined as the state of an individual animal as regards its attempts to cope with its environment (Broom, 1988). When conditions are difficult, individual animals use various methods to try to counteract those difficulties. Most animals have evolved mechanisms to cope with problems they are likely to encounter in their natural environment, such as extreme temperatures, desiccation, attack by predators or conspecifics and invasion by parasites. Many of the issues of animal welfare arise because these mechanisms become either inadequate or inappropriate when the animal is faced with a man-made change to its environment. When these mechanisms are overtaxed, the animal becomes stressed and its fitness may be affected, i.e. its life expectancy or reproductive potential is reduced. One of two approaches is often used to address these issues: the first consists of direct investigation of the behaviour and/or physiological states that may indicate the underlying stress; the second manipulates the environment, in the broadest sense, to assess how the stress may be relieved.

Welfare of Domestic Cats

Since cats are neither captives from the wild nor factory-farmed production animals, slightly different welfare criteria need to be applied, compared to those that are usually prioritized in zoos and in agriculture. Moreover, as cats vary to such a degree in their association with man, unique welfare considerations apply to different populations, which will therefore be dealt with separately.

Public attitudes to cats are considerably more sympathetic than they were several centuries ago, when acts of persecution were excused on the grounds of cats' perceived association with the devil. Cats do, however, appear to still engender both positive and negative attitudes in people, and the abuse of cats remains a welfare issue. Furthermore, among those sympathetic to cats, anthropomorphic misunderstandings about behavioural needs are common (Bradshaw and Casey, 2007), which can result in inappropriate environments or owner responses. For example, many cat owners attribute emotional capacities to cats, such as the ability to feel pride and guilt, for which there is little neurological or behavioural evidence. Owners may presume that a cat sprays urine in the house in order to 'get back at them', and interpret the cat running away when shouted at as the cat 'knowing she has done wrong', whereas such a response is likely to cause further stress to the cat and is therefore counterproductive.

Much of current scientific thinking about their welfare revolves around how cats are affected by: (i) changes in the physical space available; and (ii) whether or not other animals (especially unfamiliar humans, dogs and other cats) are encountered in that space. Fundamentally these are two aspects of a single question, concerning the extent to which cats are solitary territorial animals, in contrast to the domestic dog, which is essentially a social creature. There may be no simple answer to this, since individual cats vary considerably both qualitatively and quantitatively in the ways they adapt to changes in their physical and social environment; some of these differences can be related to the cat's lifetime experience (e.g. Dybdall *et al.*, 2007), but much variation remains to be accounted for, and it is likely that genetic differences – for example, in the shy–bold continuum – are also important.

As in other mammals, much more is known about signs that a cat is experiencing negative welfare than the converse, indicators of well-being. Negative welfare can have physical or psychological causes. The sensations of pain induce behavioural change, often expressed as sudden withdrawal or defensive aggression, although some cats appear to be able to inhibit these responses and may therefore 'suffer in silence' unless subtler signs of suffering, such as dilated pupils and lip-licking, are recognized (Robertson and Lascelles, 2010). Psychological stress is probably a much more common cause of welfare decrements, but may be more difficult to recognize. Stress responses may be acute – the 'fight-or-flight' response involving hormones from the adrenal medulla – or chronic, resulting from prolonged exposure to unavoidable stressors and mediated more by the adrenal cortex and, specifically, by the hormone cortisol. Many cats demonstrate stress by observable changes in behaviour, for example a lowered threshold for defensive aggression. Other cats respond by withdrawing, or by inhibiting behaviour, including maintenance behaviours such as grooming or eliminating. As with acute stress, indicators of chronic stress vary between individuals, but are likely to include several of the behavioural signs shown in Table 10.1. Unlike many other members of the cat family, few domestic cats respond to prolonged confinement by developing pacing stereotypies or

Table 10.1. Some behavioural indicators of stress in cats (from Casey and Bradshaw, 2005).

Type of behaviour	Examples
Inhibition of normal maintenance behaviour	Reduction in frequency of: grooming, drinking, eating, urination, defecation
General inhibition	Locomotion
Concealment	Hiding, digging in substrate
Defensive aggression	Reduced thresholds for spitting, growling, biting, clawing; redirected aggression
Conflict	Abrupt self-grooming, head-shaking, twitching
Inhibition of play	Performance of play may be an indicator of positive well-being

other kinds of repetitive behaviour, although a head-shaking stereotypy has been described, and other abnormal responses such as pica and over-grooming can develop in response to stress (see Chapter 12, this volume). Individual cats also vary in their physiological responses to stress, and the correspondences between physiological and behavioural indicators of stress are still not yet fully understood (e.g. McCobb *et al.*, 2005).

Measuring Distress from Behaviour

The behavioural signs of both acute and chronic stress have been combined together to form the 'cat stress score' (CSS), having ten levels as initially devised by Sandra McCune, and subsequently seven as modified by Kessler and Turner (see Casey and Bradshaw, 2005). Since in practice most cats score between 2 and 4, we have added an additional two half-points to the scale (Table 10.2) to improve its sensitivity. All versions of the scale are ordinal, not metric: in other words, there is no assumption that if a cat's score drops from 4 to 3 its subjective stress has changed by the same amount as that of another cat whose score reduces from 7 to 6. Individual variation in overt responses to stress also means that two cats with the same score are not necessarily suffering to the same extent. The most reliable use for such scales will be to track changes in stress in individuals or a discrete population across time, or between conditions (such as two housing types). The CSS has, for example, been used to track the adaptation of cats to different types of housing (Kessler and Turner, 1997).

Since cats may vary in their acceptance of humans and other cats, simple approach tests have been devised for both (Table 10.3), and also a more extensive feline temperament profile test has been devised to assess interactions with an unfamiliar person (Siegford *et al.*, 2003). None of these tests has yet been extensively validated.

Table 10.2. Nine-level cat stress score (CaSS9) (adapted from Kessler and Turner, 1997).

Score	Body	Belly	Legs	Tail	Head	Eyes	Pupils	Ears	Whiskers	Vocalization	Activity
1 Fully relaxed	Laid out on side or on back	Exposed, slow ventilation	Fully extended	Extended or loosely wrapped	Laid on surface with chin up or on the surface	Closed or half opened, may be blinking slowly	Normal	Half-back (normal)	Lateral (normal)	None	Sleeping or resting
2 Weakly relaxed	i[a]: laid ventrally or half on side or sitting a[a]: standing or moving, back horizontal	Exposed or not exposed, slow or normal ventilation	i: bent, hind legs may be laid out a: when standing, extended	i: extended or loosely wrapped a: up or loosely downwards	Laid on the surface or over the body, some movement	Closed, half-open or normal open	Normal	Half-back or erected to front	Lateral or forward	None	Sleeping, resting, alert or active, may be playing
2.5 As 2, except:	i: laid ventrally or sitting			i: close to body, still	Over the body, some movement						Resting, alert or active

Continued

Table 10.2. Continued.

Score	Body	Belly	Legs	Tail	Head	Eyes	Pupils	Ears	Whiskers	Vocalization	Activity
3 Weakly tense	i: laid ventrally or sitting or standing or moving, back horizontal	Not exposed, normal ventilation	i: bent a: when standing, extended	May be twitching: i: on the body or curved backwards, a: up or tense downwards	Over the body, some movement	Normal open	Normal	Half-back or erected to front or back and forward on head	Lateral or forward	Miaow or quiet	Resting awake, or actively exploring
3.5 As 3, except:				a: tense downwards		Wide open					
4 Very tense	i: laid ventrally, rolled or sitting a: standing or moving, body behind lower than in front	Not exposed, normal ventilation	i: bent a: when standing, hind legs bent, in front extended	i: close to the body a: tense downwards or curled forward, may be twitching	Over the body or pressed to body, little or no movement	Wide open or pressed together	Normal or partially dilated	Erected to front or back, or back and forward on head	Lateral or forward	Miaow, plaintive miaow or quiet	Tense ('false') sleeping, resting or alert, may be actively exploring, trying to escape
5 Fearful, stiff	i: laid ventrally or sitting a: standing or moving, body behind lower than in front	Not exposed, normal or fast ventilation	i: bent a: bent near to surface	i: close to the body a: curled forward close to the body	On the plane of the body, less or no movement	Wide open	Dilated	Partially flattened	Lateral or forward or back	Plaintive miaow, yowling, growling or quiet	Alert, may be actively trying to escape

6 Very fearful	i: laid ventrally or crouched directly on top of all paws, may be shaking. a: whole body near to ground, crawling, may be shaking	Not exposed, fast ventilation	i: bent, a: bent near to surface	i: close to the body. a: curled forward close to the body	Near to surface, motionless	Fully open	Fully dilated	Fully flattened	Back	Plaintive miaow, yowling, growling or quiet	Motionless alert or actively prowling
7 Terrorized	Crouched directly on top of all fours, shaking	Not exposed, fast ventilation	Bent	Close to the body	Lower than the body, motionless	Fully open	Fully dilated	Fully flattened back on head	Back	Plaintive miaow, yowling, growling or quiet	

[a]i (or unspecified), inactive; a, active.

Behavioural methods for probing the subjective states of cats lag behind those devised for other domestic species. It has been proposed that in the future it may be possible to use anticipatory behaviour and cognitive bias to probe cats' psychological welfare (Tami *et al.*, 2011a,b; Fig. 10.1). Similar techniques have been explored for domestic dogs: in that species, while behaviour indicating anticipation of a pleasurable event may be difficult to distinguish from behaviour indicating mild frustration (A.J. Pullen and J.W.S. Bradshaw, unpublished),

Table 10.3. Human approach test and cat approach test scores (modified from Kessler and Turner, 1999). In the human approach test, the observer approaches the cage from the front and greets the cat with 'hello cat', stands for 60 s in front of the cage touching the bars with one hand, then opens the cage for a few seconds and finally shuts it. In the cat approach test, a test cat known to remain calm in the presence of other cats is placed in a portable cage, while the cage is then covered with a cloth and placed in the enclosure containing the cat to be tested. After 4 min the cover is pulled back from the front half of the cage, and the behaviour of the cat being tested is observed for 1 min.

Score	Description of behaviour
1	Reacts in an extremely friendly way to person/cat
2	Reacts in a friendly way to person/cat
3	Turns towards person/cat, but does not approach
4	Moves away from cat or moves away and avoids any contact with person
5	Reacts in an unfriendly way to person/cat
6	Reacts in an extremely unfriendly way to person/cat

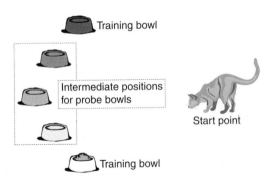

Fig. 10.1. Testing 'cognitive bias' in cats (Tami *et al.*, 2011a,b). Cats are initially trained in a spatial discrimination task, in which one location is rewarded (in the diagram, the white bowl to the cat's left, although in fact all the bowls are the same colour) while another an equal distance away is not (the darkest grey bowl). The cats are then tested for how quickly they run to a bowl placed in one of three possible intermediate 'ambiguous' locations (shown as intermediate shades of grey). Long latencies to reach ambiguous positions indicate negative cognitive bias.

cognitive bias (measured as responses to ambiguous spatial stimuli) has been successfully validated against other measures of welfare status (Mendl *et al.*, 2010).

Cats in Shelters

Animal shelters take in cats for several different reasons, each of which presents a different initial set of welfare problems. During the breeding season, a substantial proportion of the shelter population is likely to be kittens, either abandoned themselves or born to abandoned mothers. Many people seem to have little compunction in bringing unwanted litters to shelters. Provided the staff and volunteers at the shelter have sufficient time to socialize the kittens (see Chapter 9, this volume), it should be possible to home them successfully.

Adult cats can present less tractable problems. Problems with behaviour are not often offered up as the main reason for the abandonment of cats, perhaps because owners find it easier to cite allergies or changes in circumstances. However, in one survey of cats that were returned to rehoming centres as unsuitable for their adopting owners, behavioural disorders were the primary reason for over one-third (Casey *et al.*, 2009), suggesting that undesired behaviour may also lie behind many initial abandonments. Adult cats are often brought in due to changes in owner circumstances, as strays or are ferals taken from sites where they are no longer tolerated; many of the latter will be difficult or impossible to adapt to living in domestic environments. The physical as well as behavioural characteristics of these cats will influence their chance of homing. Old age or the need for ongoing medical care requiring substantial time or financial investment tend to reduce chance of homing, and chronic skin, respiratory or other conditions may also make cats unattractive as pets. Different charities vary in their policies with respect to euthanasia, but since space and finance are often at a premium in shelters, euthanasia of those cats coping least well with the shelter environment, and/or with the worst prospects of homing may become inevitable.

When first brought into shelters, cats experience high levels of stress, presumably because they are unable to predict or control their new environment. Studies suggest that both physiological and behavioural measures of stress decline as cats start to adapt to the environment, although the relative speed of adaptation varies considerably between individuals. In addition, individual cats find particular aspects of the shelter environment stressful, such as confinement, the proximity of other unfamiliar cats or handling by human attendants. Response to stressors varies between individuals, some reacting with extreme defensive behaviour when approached. Others show very little behavioural reaction, and may appear superficially to be 'settled', but the absence of an overt reaction may indicate more, rather than less, stress. Some of these cats not only do not groom, but often do not feed, urinate or defecate if caged overnight, signs of acute behavioural inhibition and therefore evidence for extreme stress. Even for a cat that adapts well to a caged environment, some of the activities that inevitably occur in shelters produce measureable stress, although there is some evidence that these affect strays less than cats abandoned by their owners (Dybdall *et al.*, 2007). These include irregular feeding and handling regimes, relocation to a new environment, travel in a cat carrier, noise (especially dogs barking), visual contact with unfamiliar cats and lack of petting from caregivers. A predictable environment, both in time and space, is likely to be particularly important for cats on entering a shelter to enable them to adapt. This includes trying to maintain a consistent olfactory environment (e.g. by using two beds and only replacing one daily) and having a minimal number of different staff members handling cats. The importance of human contact is likely to vary between cats, as some may highly value interaction and others may perceive approach as a threat. Providing opportunities for cats to show normal 'coping responses' to stressors (e.g. climbing or hiding) is also beneficial, and the simple provision of boxes can significantly reduce behavioural measures of stress (Kry and Casey, 2007).

Group housing

Cats that are already familiar with one another, for example those that have lived together amicably before being rescued, may benefit from being housed together. However, cats that have lived together do not automatically regard each other as members of the same social group (see Chapter 5, this volume), and hence care should be taken when considering group housing. Pairs of cats that have successfully avoided each other in a home environment may be highly stressed when placed in close confinement together (Fig. 10.2).

Cats should only be housed together, therefore, where there are clear signs of social bonds (e.g. allogrooming). For reasons of space and ease of care, some shelters find it necessary to house unfamiliar adult cats communally rather than singly: this can be even more problematic. When an individual cat is introduced into an existing group, it is likely to experience a degree of stress, which can come from three different sources: (i) residual stress from recent trapping and displacement from its original home range; (ii) the unfamiliarity of the physical environment into which it has been placed; and (iii) the presence of strange conspecifics – previous experience of other cats affects how much stress accrues from this (Kessler and Turner, 1999). Newly introduced cats may be aggressive towards the other occupants; on being approached they often hiss or growl, put their ears back in the defensive position and move away. All these patterns can reasonably be ascribed to social stress. They also make attempts to escape, climbing and biting the bars of the doors. Although such overt signs of behaviour often decline after the first 4 days of communal confinement, it is likely that cats remain stressed for prolonged periods. Other behavioural indicators of stress reduce more slowly, including the amount of time spent underneath the shelves, exploring the room and sitting alert (Smith *et al.*, 1994), and the lack of overt signs may occur due to behavioural inhibition or 'learned helplessness' (i.e. no behavioural strategy successfully achieves avoidance of other cats), rather than cats becoming more relaxed.

Fig. 10.2. Two cats housed together in a rehoming centre, both showing visible signs of stress.

Adult cats that have been housed together for several months or years may gradually become tolerant of conspecifics. In one study (Smith *et al.*, 1994), seven out of nine cats that had been housed together for less than 1 year were never recorded as having been in contact with another cat, whereas nine cats, of which seven had been at the shelter for over 1 year, rested in contact with another cat for part of the day. Other patterns of behaviour also differed between the most recently introduced cats and those that had been in the group for over a year: the former were the most vigilant and aggressive, while the latter were much more likely to approach other cats and to initiate mutual rubbing and bouts of grooming. The conditions under which these affiliations had been established could not be determined, and it is possible that they had arisen during some special set of circumstances. However, these data do suggest that at least some cats can eventually establish sociable relationships even when confined, relationships that are likely to be beneficial in terms of their welfare. Other cats, for reasons which are not yet clear, may never form such associations, and their welfare may therefore be best served by housing them singly.

To avoid inducing stress, group housing needs to be implemented with the cat's behavioural needs in mind. The limited repertoire of short-range social signals, especially the apparent lack of a classic submissive posture (see Chapter 5, this volume), places constraints on how quickly two cats can establish an amicable relationship, if at all. Regular removal and replacement of group members is stressful not only for the cats being added, but for the existing group members (Ottway and Hawkins, 2003). Although there is no definitive recommendation for maximum group size, 25 is generally considered the maximum for permanent groups, and fewer if group membership changes regularly (Rochlitz, 2005b). Although it is generally recommended that each cat is given a minimum of 1 m^2 of floor area, it is the surface area available above the floor in the form of shelves, platforms, walkways, windowsills and climbing posts that is more extensively used by cats than the floor itself.

Whether in group housing or penned singly, many cats seek out and make extensive use of places where they can be out of sight of other cats and/or people, depending on the personality of the cat (Barry and Crowell-Davis, 1999). The hide-and-perch box developed by the British Columbia SPCA provides both a hiding place and a vantage point, and thereby caters for two behavioural needs that are expressed to different extents in different cats (Kry and Casey, 2007). Cats housed in groups need to be monitored regularly for signs of stress, including lack of activity and interaction which may be indicative of behavioural inhibition. The risks of inducing considerable and long-term stress in cats, together with the risk of infectious disease transmission, mean that group housing is now rarely used routinely in shelters. However, it can be valuable in situations where cats need to be maintained in stable groups over longer periods, for example groups of cats testing positive for feline immunodeficiency virus.

Environmental enrichment

Traditionally, the demands of hygiene had the effect of reducing the complexity of cat pens to a bare minimum, thereby rendering it impossible for cats to express their full behavioural repertoire. Simply adding complexity ('enrichment'), especially if it is misinformed by anthropomorphism or anthropocentrism (Bradshaw and Casey, 2007), may not bring any lasting benefit to the cat unless it satisfies one of two criteria: (i) it makes it more likely that the cat will be rehomed, by influencing the behaviour of potential adopters (Wells and Hepper, 1992); and (ii) the cat's behavioural repertoire is enhanced.

It has been postulated that animate enrichments are intrinsically more valuable than inanimate, because of their greater complexity and unpredictability (Dantas-Divers et al., 2011), with the proviso that individual cats vary greatly in their tolerance of other cats and of people. Group housing (see above) may therefore reduce chronic stress in some cats, but may be aversive for others. Cats that have been socialized to humans appear to benefit, both in the short and long term, from contact with people (Casey and Bradshaw, 2008), but contact can be aversive for others.

Since the provision of extensive contact may not always be practicable, inanimate enrichments (Table 10.4) are generally easier to deploy. The use of

Table 10.4. Classification of commonly used environmental enrichments, with specific examples for cats confined in pens or indoors (modified from Ellis, 2009).

Type	Category		Examples
Animate	Intraspecific		Well-socialized feline companion
	Interspecific	Human	Handling, grooming, training
		Other	Well-socialized canine companion
Inanimate	Mobile	Toys	Balls, suspended 'prey'
		Feeding	Puzzle-feeders, hidden food
	Fixed	Substrate	Fleece, carpet, shredded paper
		Refuge	Hiding boxes
		Vantage points	Shelving, cat 'tree'
	Sensory	Visual	Window, TV
		Auditory	Human voice
		Olfactory (general)	Herbs, catnip, prey odour
		Olfactory (communicative)	Proprietary scent dispensers

boxes to provide hiding places and vantage points has already been mentioned. Interactive devices such as puzzle-feeders and suspended toys appear to provide more sustained enrichment than more passive additions such as TV screens and odour dispensers (see Ellis, 2009, for a review). It is important to consider the most appropriate enrichment for the individual needs of each cat. For example, in the period after entering a rescue shelter, cats benefit most from a highly predictive environment and the opportunity to show coping responses, such as hiding. During this period, therefore, the provision of boxes and a regular routine is likely to be more beneficial than enrichments, which cause additional complexity or unpredictability. However, once cats adapt to the shelter environment, a gradual increase in complexity and a reduction in the predictability of events may be beneficial.

Pet Cats

Since it has become increasingly common to keep cats indoors for their entire lifetimes, there is a pressing need for science to address whether this practice actually benefits their welfare, and, if so, how the indoor environment can best be tailored to their needs. Cats that are allowed to roam put themselves at risk from road accidents, getting lost, attacks by other cats and infection by viruses. Outdoor cats also impinge on the welfare of wildlife (see below). However, apart from pedigree animals, cats have not been evolutionarily selected for life in a confined space (see Jongman, 2007, for a review). Despite this pressing need, few studies have addressed the welfare of indoor cats (Bernstein, 2007), although many of the recommendations arising out of research on environmental enrichment for cats in animal shelters can also be applied to indoor pet cats.

The spatial requirements of pet cats, especially if they are neutered, are still not fully understood. House cats may range over 0.5 ha or more if given the opportunity, depending on the density of cats in their neighbourhood (see Chapter 9, this volume), but there is no compelling evidence to suggest that simply confining a cat indoors for its entire lifetime causes it to suffer. It appears likely that the spatial requirements of domestic cats are highly flexible, and that poor welfare only results when an abrupt change occurs in the area available. This could be a reduction, as when a feral cat is trapped and confined in a shelter, or an increase, for example if an indoor cat is abandoned outdoors.

One of the most common causes for behavioural signs of stress in domestic cats is living in close proximity with other cats that they do not regard as socially compatible, whether within the same household or in the neighbourhood. Owners commonly misunderstand the limitations of social signalling in their cats (see Chapter 5, this volume) and hence do not provide sufficient opportunities for the cats to avoid one another, and this can lead to a range of undesired behaviours (see Chapter 11, this volume). The welfare of pet cats can also be compromised by interactions with owners, particularly where tolerance of close contact is limited in cats with limited early socialization with people, or where interactions are particularly inconsistent.

Feral and Stray Cats

Public reaction is variable towards cats living wild, either strays that have abandoned or been abandoned by their owners, or ferals that have never been owned. Those who are distressed by the sight of these cats and their kittens, which often appear to be unhealthy and ill fed, may react either by trying to provide food and care or by demanding that they be destroyed. Other complaints relating to feral cats include:

1. Fouling of gardens and communal areas, giving offence by sight and smell.
2. Nocturnal fighting and caterwauling.
3. Finding of corpses, and diseased and dying cats and kittens.
4. Attacks on pets and people.
5. Entering homes uninvited.
6. Fleas.
7. Being a 'health risk' to pets, children and babies (potential health hazards include ringworm, toxoplasmosis, toxocariasis and rabies).
8. Killing and scaring of birds, small mammals and ornamental fish.
9. Digging up gardens.

While some of these problems may be more widespread than others, many are genuine conflicts with the interests of people living near to feral colonies, and many have considerable behavioural components. Although the most obvious solution, and one that is often adopted, is to eradicate feral colonies by poisoning, shooting or trapping (see Robertson, 2008, for a review of control measures), this often has to be repeated on a regular basis because new cats move in to exploit the resource that attracted the original group. While strictly speaking the painless destruction of an animal is an ethical rather than a welfare issue, cat lovers are obviously distressed by such procedures. The 'trap–neuter–return' (TNR) procedure has been widely adopted as an apparently humane alternative. As many of the cats as possible are trapped; old and incurably diseased animals are euthanized, socialized adults and socializable kittens are homed and the remaining adults are neutered and returned to the original site. Marking – for example, by removing the tip of the left ear at the same time as sterilization – prevents the same cat from being repeatedly presented for neutering. Continued neutering of entire immigrants is needed, otherwise breeding will quickly resume.

The behavioural effects of TNR are consistent with the known effects of neutering: both female–female and male–male aggression are reduced and the general stress levels of the cats appear to be reduced (Finkler and Terkel, 2010; Finkler *et al.*, 2011). Although the cohesiveness of the original group appears to be unaffected by neutering (see Chapter 9, this volume) it is unclear whether neutered colonies defend their territories as successfully as do breeding colonies (Brown and Bradshaw, 1996), raising the possibility that whole colonies may be more easily displaced post-TNR.

TNR is not universally supported, even among the veterinary profession: for example, Jessup (2004) refers to it as 'trap, neuter, re-abandon' and questions its legality in the USA. Many wildlife conservationists favour extermination, because feral cats continue to hunt even after neutering (e.g. Longcore *et al.*, 2009).

Undesired Behaviour in the Domestic Cat

Introduction

In this chapter we examine behaviour that cat owners may find problematic or undesirable. Such behaviour has an impact on the cat–owner bond, can result in relinquishment or even euthanasia of cats and can also indicate compromised welfare in the cats themselves. Despite this, relatively little research has focused on the epidemiology, aetiology, prevention or treatment of such behaviour in cats, and much of the published literature is anecdotal, based on individual case reports or derived from opinion-based sources.

Although terms such as 'behaviour problem' and 'undesired behaviour' are widely used in the literature, it is difficult to find a clear definition of what such terms mean, or a consistent idea of which types or presentations of behaviour are included. Since a 'behaviour problem' must be a behaviour that an owner finds 'problematic', this will clearly vary with each individual owner's subjective interpretation of what is acceptable. For example, whereas one owner may find intermittent aggressive behaviour between cats in a household acceptable, another may interpret this behaviour as a problem. Indeed, further complications for such definitions occur where owners interpret common behavioural signs shown by their cats in different ways: for example the same behaviour may be perceived as 'aggression' by some but regarded as 'play' by others. Our first problem in understanding undesired behaviour, therefore, is to consider how such behaviour may be classified, and the extent to which owner perception of 'undesirability' is an important consideration in such definitions.

The other potentially perplexing aspect of defining undesired behaviour is the extent to which such behaviour is considered to lie within the normal repertoire of

© J.W.S. Bradshaw, R.A. Casey and S.L. Brown 2012. *The Behaviour of the Domestic Cat*, 2nd Edn (J.W.S. Bradshaw *et al.*)

the cat. Indeed, behaviour problems are sometimes divided in the literature into those that arise from behaviour that is essentially 'normal' or 'adaptive' for the species, and those that appear to be 'abnormal' or 'maladaptive' (Borchelt and Voith, 1982). The former includes those instances where behaviour is likely to be shown by cats in a manner compatible with that seen in a 'natural' environment, or which may occur in other contexts consequent to modulation by the processes of associative learning. For a domesticated species, such as the cat, it is perhaps difficult to define what a 'normal' environment may be, and in practice comparisons are generally made with cats in a free-ranging or feral environment. Abnormal behaviours are generally considered to include those with a clearly medical origin, or those where the behavioural signs displayed are not generally apparent in that species.

Differentiating 'normal' or 'abnormal' behaviour, however, is not as clear as might be expected. Some presentations do fall clearly within one or other of these categories: a cat that urine sprays in response to enforced close contact with another cat that is not part of its social group can be described as showing an adaptive response, even though this may be undesirable to an owner. Equally, an extreme behavioural response that occurs entirely independently of environmental stimuli, for example from a partial seizure focus within the limbic system, is clearly 'abnormal'. However, there are many cases that fall between these two extremes. Many animals show species-specific behaviour in response to an aversive event, but may either appear to react at an unexpectedly low threshold, show a higher degree of response than would normally be expected at first exposure or generalize a behaviour between contexts more rapidly than one might expect through learning. There may be many reasons for these apparently 'abnormal' characteristics of 'normal' behaviours, including: (i) medical conditions that influence the threshold of responses (see Chapter 12, this volume); (ii) other environmental causes of arousal occurring concomitantly; (iii) genetic factors such as personality characteristics; (iv) developmental causes of increased reactivity; (v) epigenetic changes caused by chronic stress; or (vi) a whole host of other factors. Other cases might present with behaviour that appears on first examination to be completely abnormal, but for which a clear history emerges, revealing how the cat has learnt the behaviour through a series of unique experiences over time. It is the complex interweaving of genetic, developmental, environmental and internal factors that can make the interpretation of individual behaviour cases both fascinating and challenging. The involvement of multiple, interacting factors in each case also means that those who advise owners on the treatment and management for such cases should have both demonstrable knowledge and clinical skill in feline behaviour before attempting to treat cases. The overlapping roles of learning and medical disease also mean that non-veterinary clinicians should work closely with members of the veterinary profession.

Proportion of Cats Showing Undesired Behaviours

There is limited evidence as to the proportion of domestic cats showing undesired behaviour. However, the number of cats referred to specialist centres for behavioural advice is suggested to be increasing: whether this change is as a result of increased owner awareness that behaviour therapy is an option for cats, or due to changes in the actual number of behaviour problems that are developing, is unclear (Heath, 2005). The former is possible, since cats appear to be becoming increasingly popular pets, with a recent survey estimating over 10.5 million cats in the UK (Murray *et al.*, 2010). In addition, the role of the cat in households also appears to have altered over time, from one where the cat was rather more peripheral to the family, possibly kept for rodent control, to one where the cat plays an important social role within the family and can provide considerable emotional support for some owners (Bradshaw and Limond, 1997). This may result in owners investing more time and financial resource in addressing undesired behaviours.

These same factors, however, may also account for an increased proportion of cats developing undesired behaviour. The increasing number of cats inevitably leads to higher local population densities. The stress that many cats experience when living in close proximity with unfamiliar cats (see Chapter 8, this volume), particularly where local populations are in flux, may account for increased numbers of undesired behaviours. More cats are also kept in multi-cat households, with Murray *et al.* (2010) reporting that 42% of cat owners had two or more cats, and 2% having between six and 12 cats; such numbers are likely to generate problems associated with social stress, as discussed further below. The changing role of cats within families may also be an important factor in increased behaviour problems: the inclination of owners to show intense and close bouts of interaction with cats can be at odds with the natural social behaviour of cats, and the resulting miscommunication may lead to the development of behaviours that owners find undesirable.

The Association of Pet Behaviour Counsellors (APBC) in the UK produces an annual review of data about the cases that have been referred to members from veterinary surgeons. Since rather few members of this organization specialize in feline behaviour, the data are limited in their application, but they do provide some insight into the types of behaviours for which cat owners seek expert help. In the 2005 review of cases, information was collated on 65 feline cases, of which 30 were male and 35 female. The presenting problem for 28% of the cats was indoor marking (including urine spraying, squat marking, middening and scratching) and 11% presented with inappropriate toileting problems; 22% displayed aggression to other cats, of which 86% was directed towards other cats within the household; 11% showed some kind of fearful or 'phobic' response; 10% of cases presented with aggression towards people, 7% with behaviours caused by medical disorders and 5% with 'bonding problems' (presumably meaning cats showing a limited tolerance of interaction with people) (APBC, 2005). Data collected over a 6-month period in a referral clinic (Fig. 11.1) similarly showed that urine spraying, inappropriate elimination, aggression between household cats and aggression directed towards owners were the most common reasons for which owners sought help.

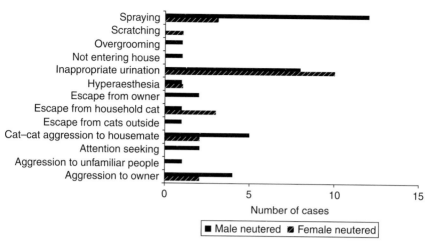

Fig. 11.1. Numbers of cats, by gender, referred to a specialist centre over a 6-month period in 2000 (*N* = 61).

Surprisingly little research has been done to investigate the prevalence of behaviour problems in the general cat population. Bradshaw *et al.* (2000a) conducted a door-to-door survey of 90 households in two areas of the UK, one rural and the other suburban. Answers from 15 male and 75 female respondents were acquired regarding all cats present in the households, a total of 161 cats. As shown in Fig. 11.2, not only was there a surprisingly high number of undesired behaviours displayed by cats in this study, but the frequency of the various types of problem was very different from that described above in a referred population. A survey of 109 owners visiting a first-opinion veterinary practice similarly showed a higher proportion of behaviour such as avoidance of other cats, scratching and avoidance of visitors, compared with the common problems seen on referral (Fig. 11.3).

The difference between numbers and types of behaviour problems in the referral population and in the general population surveys suggest that the cats seen by referral practitioners are not only the 'tip of the iceberg' in terms of numbers of cats with behaviour problems, but also that owners tend to seek help for particular types of behaviour. These appear to be those behaviours that are more likely to impinge on the lifestyle or environment of owners, such as urine spraying, inappropriate urination and aggression towards people or other cats in the household. Confirming this, a general prevalence survey found a high proportion of problems that had less impact on owners, such as avoidance of unfamiliar people or other cats (Casey and Bradshaw, 2001).

Other evidence for the extent to which undesired behaviours in cats influence the cat–owner bond comes from reasons for relinquishment of cats to rescue organizations. There is considerable inconsistency in the proportion of cats reported to be relinquished for apparent behaviour problems in these studies, ranging from about 8% to over 33%, probably influenced not only by methodological differences in data collection but also by owner perception of what may be an 'acceptable'

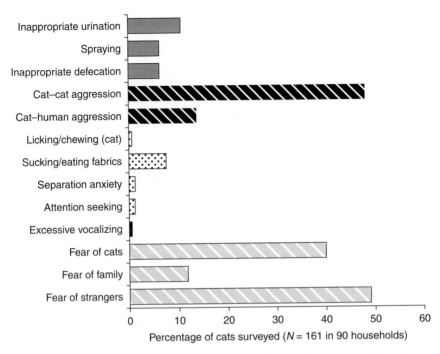

Fig. 11.2. Prevalence of 'behaviour problems' in the general cat population (from Bradshaw *et al.*, 2000a).

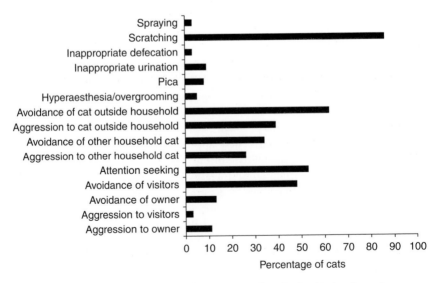

Fig. 11.3. Percentage of cats showing a range of undesired behaviours in a survey of cat owners visiting a first-opinion veterinary practice (*N* = 109) (from R. Casey, unpublished data).

reason to abandon a cat. In the UK, over a 12-month period, only 7% of cats were relinquished to Cats Protection primarily due to undesired behaviours (according to owner reports), although a much higher proportion (38%) were returned to centres after homing because of undesired behaviour (Casey *et al.*, 2009).

Classification of Undesired Behaviour

Consistent descriptions and classifications of undesired behaviours are essential for those working in cat behaviour research or clinical practice. However, classification of undesired behaviours has taken a number of different approaches in the literature. For example, behaviour problems in cats have been defined according to: (i) what form the behaviour takes (e.g. 'excessive vocalization'); (ii) the context in which the behaviour occurs (e.g. 'territorial aggression'); (iii) the target for the behaviour (e.g. 'intraspecific aggression'); or (iv) the likely motivation of the behaviour (e.g. 'redirected aggression'). All of these approaches have potential drawbacks. For example, describing an observed behaviour, context or target gives no indication as to potential motivation. This can be misleading, since a behavioural presentation can arise via a number of different routes. For example, an excessively vocalizing cat could be showing this behaviour for one or more of several reasons, including gaining owner attention, in response to anxiety, or as a consequence of hyperthyroidism or age-related changes. A diagnostic category combining these together could lead to prescriptive approaches to treatment that are unlikely to be appropriate for all cases.

The main drawback of defining behaviour according to motivation is that this approach requires a degree of subjective interpretation, gleaned from historical, observational and contextual information. Because of the interpretational nature of this type of classification, variation may occur between authors, for example relating to differences in preferred theories of aetiology within different cultures. With the current limited knowledge of the aetiology of behavioural conditions, these diagnoses must be regarded as hypothetical constructs that should be subject to modification and development as science progresses and knowledge increases (Sheppard and Mills, 2003). One salient example of difficulties arising from this type of classification system is the differences of opinion in the use of a hierarchical framework when interpreting interactions between groups of cats (see Chapter 8, this volume). Motivational categories of behaviour are, however, likely to be more closely related to treatment protocols.

Factors Influencing Development of Undesired Behaviours

Although undesired behaviours are often considered separately, the principles underlying the development of such behaviours are no different from those

influencing behaviours that are acceptable for owners. In interpreting undesired behaviour it is therefore important to consider the normal behaviour or
ethology of cats, together with the individual learning opportunities that have
influenced the development of specific responses in particular circumstances.
Most of the undesired behaviours that we discuss in this section are essentially
'normal' responses of members of this species to the environment in which
they find themselves. However, owners often seek help because behaviours are
incompatible with their lifestyles, are elicited by inappropriate stimuli, occur
in an inappropriate context or are shown at such intensity that they become a
nuisance. In many cases, such behaviour has started as a normal response and
becomes inadvertently reinforced to the point that it becomes unacceptable.
Many behaviour problems also arise as a result of an owner's lack of understanding of the natural behaviour of their pet. For example, many owners do
not appreciate the amount of mental and physical stimulation that cats require,
particularly when housed in an indoor environment, and the occurrence of
many behaviour problems is at least partly influenced by their pet's lack of
opportunity to show normal species-specific behaviour. Apart from physiological and pathological factors, which are addressed separately in Chapter 12, this
volume, the main factors influencing the development of undesired behaviours
in cats are: (i) the limitations of intraspecific social behaviour in the species; (ii)
relative differences in socialization experience; (iii) breed and individual differences; (iv) ability to show natural behavioural responses; and (v) individual learnt
experiences. These are separately addressed in the sections below.

Social interaction between cats

The unique origin of the domestic cat, derived from the essentially asocial *Felis s.
lybica*, and subsequently adapted to live in social groups, has important impacts
on the ability of *F. catus* to live successfully in the domestic environment. Their
asocial ancestry has resulted in a much more limited visual signalling repertoire
than that of species derived from cooperative hunters, such as the domestic dog
(see Chapter 5, this volume). Complex visual signalling evolved as an important
element in enabling group cooperation in naturally social species, but was not a
necessity for *F. s. lybica* where adult individuals maintained a distance that obvi-
ated the need for signals suitable for face-to-face encounters. Whereas for the
ancestors of the domestic dog, social success depended on the ability of group
members to display changes in emotional state such that other group members
could adapt their behaviour accordingly, this ability was not important in the
evolution of the cat.

The limitations of visual signalling in cats appear to have been a factor in
restricting the types of social groupings to which the species can successfully
adapt. Feral or farm colonies are made up of cooperative social groups of females,
based on the concentration of food resources, and providing advantages of cooperative rearing of kittens (see Chapter 8, this volume). These groups work well,

despite the relatively poor ability of cats to display changes in emotional state using subtle visual signals, for two broad reasons. The first is that competition between individuals is minimized by the group size closely matching resource availability, and hunting behaviour remaining a solitary exercise. The second is that members of social groups are predominantly familial. Individuals are very familiar with each other, often developing together since birth. Furthermore, social bonds are repeatedly reinforced with affiliative tactile behaviours and the swapping of scent signals through rubbing and grooming. This familiarity enables group members to 'know' each other well, in the sense that they can effectively predict how other group members are likely to respond in different circumstances, making the ability to judge responses through interpretation of visual signals less important. Encounters with cats from other colonies are unlikely to result in affiliative responses; rather, such interactions generally result in strategies that avoid any contact at all, or involve overt signs of aggression. Studies of these natural grouping of free-ranging cats therefore suggest that the response of cats to conspecifics is highly dependent on their relative perception of others as a 'group member' or 'non-group member'.

Given the limitations of social compatibility for cats, it is perhaps unsurprising that the diversity of social situations in which pet cats find themselves is the most common cause of undesired behaviours. In areas of high human population density, cats may live in close proximity with a number of cats which they do not perceive as part of the same social group. This can lead to high levels of vigilance and anxiety as cats attempt to occupy home ranges while at the same time avoiding contact with other cats. Predicting the temporal and spatial activity of other cats, for example through the monitoring of scent signals, can sometimes enable cats to occupy overlapping territories but not come into contact. However, where altered patterns of activity are caused by new cats arriving in an area, owner activities such as holidays or altering cats' access outside, this can have an impact on cats across a considerable area. The greatest impact is probably caused by cats that maintain a large territory size (e.g. unneutered or late-neutered toms), since they will come into contact with other cats over a wide area.

Behaviours arising from conflict between neighbourhood cats

Conflict between cats in a neighbourhood can have a number of consequences for the behaviour of individuals. Most obvious is the occurrence of aggression between cats: although owners may hear evidence of cats fighting, the first sign of a problem is often the abscesses occurring as a result of cat bites. The limited ability of cats to show appeasement behaviour means that encounters between cats often result in overt aggression, chasing and biting. However, other behavioural consequences are also common: cats that are anxious about contact with other cats cannot predict the activity of other cats; cats that have had previous negative encounters with other cats may alter their patterns of outdoor activity. This can consist of reducing time outside, the avoidance of particular areas or reduced overall area of range. The consequences of such alterations will vary with other factors, but can include, for example, increased amounts of time seeking interaction with owners. Such cats may seek attention through excessive vocalization, or show abnormal play/predatory responses towards owners. Alternatively, avoidance of contact with neighbourhood cats may result in some cats no longer leaving the safety of the home, or only dashing out for short periods. Cats that are anxious about the activity of other cats in close proximity to the home may spend considerable amounts of time showing vigilant behaviour, such as 'checking' out of windows, or looking out through clear cat doors. They may only venture outside when owners go into the garden, or prefer to leave the house through an open door or window to avoid the uncertainty associated with going outside through a cat door with limited visual access to the outside area.

Reluctance to leave the safety of the house due to the presence of other cats can also lead to the development of inappropriate elimination. Where a cat's normal toileting location is outside, reluctance to leave the house is a common reason for seeking an alternative toileting location inside. Cats are vulnerable when toileting, and insecurity about access to a usual toileting site is one reason for cats to alter their toileting location (Table 11.1). Hence, if a cat normally toilets outside, anxiety about neighbourhood cats will commonly lead to a shift to toileting inside; where a litter tray is not provided this will be on other surfaces, and perceived as undesirable by owners.

Urine spraying is another consequence of social incompatibility between neighbourhood cats. As discussed in Chapter 5, urine spraying is an olfactory signal that is considered to serve a number of functions. In entire cats, these signals appear to have a sexual signalling function. However, urine spraying also commonly occurs in neutered domestic cats. Although historically often described as a 'territorial' behaviour, urine spraying rarely occurs at the periphery of a cat's range. Rather, spray marks usually occur in locations that are 'socially significant' (Herron, 2010). For example, the most commonly sprayed items include furniture and walls or windows near to visual access outdoors (Pryor *et al.*, 2001), which are areas where conflict with other cats is likely to occur, or where the activity of other cats may be observed. Although cats 'check' both their own scent marks and those of others, they rarely 'over-mark'

Table 11.1. Characteristics of suitable toileting sites for cats.

Features of a suitable toileting site for a cat
Suitable material for burying
Quiet location hidden away
Location away from the threat of other cats that are not perceived as part of the same social group
Some cats prefer a separate location for defecation and urination
Litter matches the cat's substrate preference
Location away from feeding site, but not so far from the core area as to feel unsafe

the scent of other cats with their own (Hart, 1974), also suggesting that this behaviour is not purely territorial. The rate of 'checking' tends to decline with the age of the scent mark, such that fresh urine is most interesting to cats, and interest decreases as the scent fades (De Boer, 1977b). As time since deposition appears to be important to cats, it seems likely that one important function of these marks is to enable cats to avoid each other when they co-occupy the same area (Cooper, 1997) in order to avoid overt aggression.

Since cats will urine spray in social isolation, it is also possible that these marks have a function in providing information to the marking cat itself. Leaving a mark in locations associated with perceived threat may provide the animal with a greater degree of predictability and control over its environment, as it enables it to 'know' areas where there is potential danger, and vigilance is needed. Hence, a spray mark on the cat flap would alert a cat to be cautious in that area if conflict with other cats had previously occurred at that location.

Cats that have their activity inhibited by other cats in the neighbourhood may also show other behavioural signs associated with stress, such as over-grooming. This is particularly likely to be the case where other cats actually enter the house, or the environment is such that the resident cat cannot predict or avoid encounters with other cats perceived as threatening.

Behaviours occurring in multi-cat households

Where multiple cats live in the same household, similar issues arise among them as those caused by social incompatibility between cats in a neighbour-hood. Whether undesired behaviours arise in multi-cat households depends on the extent to which cats perceive each other to be members of the same social group, and also on the extent to which they can access resources independently. As with free-ranging cats, individuals in social groups generally show high levels of affiliative behaviours (such as allogrooming and allorubbing) and will choose to be in close proximity to each other, for example sleeping in contact with one another (Table 11.2). Also consistent with free-ranging cats, individuals that are siblings or that have developed together as kittens are more likely to form social bonds than cats introduced as adults (see Chapter 8, this volume).

Where two or more cats within a household form separate social groups, they will generally establish different core areas in different parts of the house, and often essentially live separate lives. This may be achieved by active aggression or withdrawal on encountering the other cat, but more often occurs through establishing routines whereby actual encounters are minimized (Table 11.2). In such cases, even where there is no active aggression, the activity of each cat may be inhibited by the others. For example, one cat in the household may be unwilling to pass through a doorway when another cat is nearby, or it may not enter through a cat flap when the other cat is in the vicinity. Owners can often be oblivious to these patterns of avoidance. This is particularly the case where incompatible cats occupy the same space in order to access essential

Table 11.2. Behavioural signs shown by cats in multi-cat households indicative of social groupings.

Indicators	Behavioural signs shown between dyads of cats within a household
Indicators that cats are likely to be in the same social group	Allogrooming Allorubbing Sleeping touching each other Interactive play behaviour initiated by either cat
Indicators that cats perceive each other as being in different social groups	Chasing/running away Hissing, spitting or batting at each other Avoiding contact or hiding Displacement of one cat by another Choosing to occupy different spaces (although may come together to access valued resources, such as at feeding time) Blocking movement (e.g. one cat sitting on stairs stops the other coming down) Inhibiting movement or activity (e.g. one cat remains immobile while the other present) Tense facial or body posture during contact Vigilance whilst other cat present (e.g. watching the other cat) Cats engage in play or interaction with owner separately

resources. For example, owners will often provide food for all cats in a household in the same room at the same time. In order to obtain sufficient nutrition, cats may have to eat close to other cats, although they would not choose to do so in other circumstances. Often, signs of anxiety shown by cats when forced to eat in close proximity are often not appreciated by owners. This may include individual cats bolting food very fast, or eating excessive amounts at a time to avoid the necessity of returning frequently to the feeding area, factors potentially important in the development of obesity. Other examples of unwilling proximity may occur where cats are attracted to limited sources of heat, or where multiple cats in a household value human contact and may sit either side of an owner in order to achieve attention, while also attempting to avoid direct contact with each other.

The problems for cats of living in a household with other cats not perceived to be in the same social group are often exacerbated by their owners' tendency to cluster other important resources together. In addition to feeding cats together, owners often provide litter trays, water bowls and entrance/

exit points to the outside in single locations. The combination of limited resource access and avoiding other cats in the household commonly leads to a range of undesired behaviours. For example, a cat that has a core area upstairs in a household will have restricted access to important resources if these are all downstairs, in the core area of another cat. One common consequence might be the 'upstairs cat' starting to eliminate upstairs if access to a downstairs toileting location becomes limited. In some cases the toileting behaviour is precipitated by some other change, such as a change in the routine of the 'downstairs cat'. For example, the 'upstairs cat' may learn that it is safe to go downstairs to use the toilet first thing in the morning and last thing at night, because these are times when the 'downstairs cat' is out hunting. However, if another cat moves into the neighbourhood and inhibits the outside activities of the 'downstairs cat', this may induce inappropriate toileting in the cat living upstairs. Equally, if the cat living upstairs is in the early stages of hyperthyroidism, it will start to drink more, and hence need to urinate more than the previous twice daily, leading to the onset of inappropriate elimination upstairs.

As with conflict between cats in the neighbourhood, anxiety arising due to incompatible cats living in the same house can lead to various other undesired behaviours. One common consequence is the occurrence of urine spraying, with urine marks located in significant areas such as where cats need to pass each other in narrow passageways in order to access important resources. As described for neighbouring cats, these signals may have both a communicative function to avoid direct confrontation and enable marking cats themselves to subsequently identify areas where potential conflict may occur.

Behavioural signs of chronic stress also occur. These may include cats spending prolonged periods of time hiding to avoid contact with other cats, or showing abnormal responses such as overgrooming. In addition, some cats move away from households where contact with other cats is stressful, for example predominantly living outside or moving to live in other households.

Because cats do not appear to be motivated to maintain social bonds in the same way as other species, the breakdown of established relationships often results in a permanent split of cats into separate social groups. Since the recognition of other individuals as group members is likely to be at least partially through their odour profiles, maintained by affiliative rubbing and grooming, situations that alter the scent of a group member can result in the breaking of social bonds. Should one cat from a social group leave a household for a period, for example if it is hospitalized, even siblings may fail to recognize that cat on its return. In extreme cases, this can lead to the complete breakdown of the relationship, because the initial aggressive response leads to a reciprocal response from the home-comer. It is therefore sensible for cat owners to rub cats returning from a trip to the vets with a towel or piece of bedding that is covered in the 'group scent', prior to reintroduction, or to actively 'swap scent' by stroking one cat and then the other (Crowell-Davis *et al.*, 1997).

Relative socialization experience

The importance of early sensitive periods for learning about both conspecifics and people has been discussed in Chapters 4 and 9, respectively. Experiences during the weeks immediately after a kitten is born appear to have a profound effect on the occurrence of fear-associated behaviours later in life (Casey and Bradshaw, 2008). Most of the evidence about socialization relates to the extent to which kittens need to experience people in order to accept contact with humans as adults. Where contact with people before 8 weeks of age has been limited or restricted to people of a particular gender or age category, anxiety about people is more likely to occur subsequently.

Probably one of the most common behavioural consequences of limited socialization experience is the avoidance of unfamiliar people. Although rarely seen in the referral behaviour clinic, for reasons discussed earlier, avoiding contact with visitors appears to be particularly prevalent in the cat population. Many cats are reported to 'disappear' when visitors come to the house, but very few owners perceive this to be a problem. Cats that are fearful of their owners are presented as clinical cases more frequently. These are often cats that have had little or no experience of humans in their early sensitive period for learning because, for example, they were feral or farm kittens.

Cats may also display aggressive behaviour towards people as a result of limited socialization. Aggression occurs as an alternative strategy to avoidance when the cat is attempting to prevent an anticipated negative outcome. In general terms, cats tend to avoid threatening stimuli by withdrawing, hiding or climbing. However, aggression is used as a defensive strategy when other behaviours are not successful. For example, this might happen where owners 'pursue' cats that are frightened, and try to interact with them. Once a cat has learned that aggression is an effective strategy to keep people away it will become more confident in the expression of this behaviour. Over repeated learning opportunities, therefore, cats may develop apparently 'offensive' aggression, even launching themselves at people, or showing the behaviour immediately on perceiving a particular person. It is particularly important, therefore, that opportunities for withdrawal are provided for cats housed in confined spaces, such as rescue centres (Kry and Casey, 2007), to prevent the development of aggression in this context.

Although there is much less evidence as to how social tolerance for other cats may be learned during the sensitive period, it is likely that learning about conspecifics occurs in parallel with learning about people. Differences in the relative tolerance of individual cats for other cats may therefore be modified by early experience: this is an area where further research is important, given the importance of social stress for both the occurrence of undesired behaviour and the welfare of domestic cats. As discussed with respect to interaction with people, ongoing social experience of other cats may modify perceptions formed during the sensitive period. For example, a cat that is repeatedly attacked by another cat may subsequently feel even more threatened by the proximity of other cats.

A further important consequence of the early environment is the formation of the kitten's preferences for toileting location. A preference for substrates
on which to toilet forms in the first weeks of life. This occurs as kittens form
an association between the act of toileting and the material under their feet at
the time. As with other associations made during this period of development,
this will often influence toileting behaviour throughout life. For example, where
breeders use a particular type of litter material, kittens may not recognize other
substrates as 'toilets' after homing. It is probably also for this reason that hand-
reared kittens sometimes have a propensity for eliminating on soft furnishings.
Since their carers tend to hold them in a towel whilst stimulating the urogeni-
tal reflex (emulating the mother's usual behaviour), some appear to form an
association between toileting and soft materials. Although substrate preferences
formed during development can be strong, preferences can also change in the
adult cat. Generally this is desirable, as it enables owners to gradually change
from one litter material to another if they wish to. However, if a cat has persistently toileted in an inappropriate location, such as a carpet, over a prolonged
period, it can form a preference for this surface, making resolution of the problem more difficult.

Ability to show a normal behavioural repertoire

Another important factor in the development of undesired behaviour is the
extent to which cats are able to show a full normal behavioural repertoire. The
normal ethology of the cat and its species-specific behaviours should therefore
be considered as potentially relevant in individual cases where natural behavioural opportunities are restricted. The extent to which cats tolerate restrictions
in activity, social interaction and predatory behaviour appears to vary considerably between individuals, and is likely to be influenced both by personality
characteristics (see Chapter 9, this volume) as well as 'expectations' derived from
previous experiences (see Chapter 3, this volume). For example, cats appear
to vary considerably in the extent to which behaviour is altered by indoor-
only housing, although anecdotal evidence suggests that cats that were previously active outside are more likely to show frustration-associated responses to
spatial restriction. However, tolerance of spatial restriction will also vary with
the extent to which owners are able to enrich the indoor environment – for
example, through the provision of play to simulate hunting activity, feeding
enrichment and opportunities for climbing utilizing a three-dimensional space
(Ellis, 2009).

The ability to display predatory behaviour, or to direct predatory-type
responses into play, appears to be particularly important for the cat. Restricted
ability to show this type of behaviour can result in frustration, and thus to
undesired behaviours including human-directed aggression. Inappropriate pred-
atory/play aggression towards owners often first develops where these behaviour
patterns are misdirected in the kitten. Play behaviour in kittens is important in

the development of the motor responses needed for predatory behaviour (see Chapter 7, this volume). In a feral situation, this 'practising' is initially directed at inanimate objects, but is later directed by the queen towards prey items that she brings back to the nest site (Kitchener, 1991). The kittens therefore learn the appropriate conditioned cues that stimulate these behaviours. In the domestic environment, owners are frequently tempted to play with their kittens by, for example, wiggling their fingers or moving their feet around under a duvet. While this is relatively harmless with a kitten, it can lead to inappropriate play/predatory aggression being directed towards hands or feet once the cat matures to an adult. In these cases, cats often 'ambush' their owners as they walk past by rushing out from behind furniture and grabbing their feet or legs, or swiping at them as they walk past. Once this type of behaviour starts it is often reinforced by the response of the owner – shrieking, pulling arms away or running away tend to reinforce the response, just as movement and squealing of a prey item would encourage further attack.

The other important aspect of the normal requirements of cats that may lead to undesired behaviour is patterns of toileting. Eliminating away from normal toileting sites commonly occurs where cats cannot easily access these sites, as discussed earlier. However, toileting behaviour can also change where toileting locations no longer fulfil the cat's requirements of toileting locations, such as a suitable hidden location and litter substrate (Table 11.1). For example, since cats are generally fastidious in their eliminating habit, a litter tray that is cleaned infrequently is a possible reason for choosing an alternative location (Herron, 2010). However, since they are attracted to a toileting site partly by olfactory cues, over-zealous cleaning, particularly with a strong-smelling disinfectant, can also cause problems.

Learnt experience

Learnt experience plays an important role in the development of almost all undesired responses. For example, although relevant socialization experience may lead to a general acceptability of social contact with people, it is individual learning experience on first contact that determines how an individual cat responds to any perceived threat from a specific human contact. Furthermore, learning opportunities throughout life constantly modify how a cat responds to various aspects of its social and physical environment. Hence although first experiences strongly influence ongoing responses, behaviour can be subsequently modified. Indeed, the ability of cats to learn new associations forms the basis of behaviour modification programmes. In many cases, relative socialization experience is reinforced by subsequent experience – for example, a cat that is wary of people due to limited socialization is likely to avoid people who approach it. Where this strategy successfully avoids the perceived threat, the response will be reinforced, and hence more likely to be used on subsequent encounters (see Chapter 3, this volume). In some cases, however, fear responses occur entirely as a result of specific aversive experiences. These behaviours in cats become reinforced as they are successful at avoiding the perceived aversive event. Through the normal processes of associative learning, cats will also become gradually more sensitized to aversive events, such that they respond at a lower threshold of stimulus, and will also generalize an avoidance response to similar stimuli.

In addition to the avoidance response, cats that value human attention will commonly learn behaviours that are successful at achieving this. For example, a cat that values human attention may learn that walking along a shelf and 'wobbling' valued ornaments is a very effective way to achieve interaction with owners who were otherwise watching the TV.

Learning is also important in understanding how the behaviour of individual cats may develop over time. Urine spraying in cats often first develops as a response to the proximity of socially incompatible cats, as discussed earlier. However, once established, the occurrence of the behaviour can be modified through learning. For example, the frequency and location of urine-spraying behaviour can alter depending on how owners respond to the behaviour, and how cats perceive this response. Because owners find the behaviour undesirable, many cats that urine spray are punished, for example by being shouted at or chased out of the room. Most cats find this response aversive, and will learn to leave scent marks only when owners are not present. However, other consequences are possible from this response by owners. Some cats may become very anxious about the change in the behaviour of their owner: since they do not understand the reason for the punishment, they perceive only that their owner is behaving strangely towards them. Since spray marking is used to 'identify' contexts in which vigilance is required, cats will sometimes use the behaviour to identify owners as something of which to be wary, and urine spray directly on their owners, or on items that smell of their owners. Unfortunately many owners anthropomorphically assume that this behaviour is 'vindictive' or 'spiteful', thereby leading to an increase in punishment, and hence anxiety in the cat, and so the rate of spray marking spirals upwards. A further potential consequence of owners punishing their cat for urine spraying is where cats do not perceive the owner's response as punishing, but rather reinforcing. This occurs occasionally in cats that are highly motivated to achieve owner attention, and in such cases urine spraying may develop as a response that occurs in front of owners in order to achieve a response from them.

The manner in which cats perceive the consequences of their own actions, and behavioural strategies learnt to resolve situations of perceived threat or achieve desired goals, are therefore important factors in modulating how individual cats respond in different circumstances.

Physiological and Pathological Causes of Behavioural Change

12

Introduction

Overt behaviour is the consequence of a cat perceiving some change in its environment, evaluating this change, deciding on an appropriate response and the response being generated through the motor systems of the brain to the elements of the skeletal system that control activity. Hence, although behaviour occurs as a consequence of changes in the external environment, the responses generated also depend on internal variations in the processing of information. These processes are susceptible to alteration due not only to normal physiological variations but also to pathological changes. Interpretation of behaviour, therefore, requires an understanding of how the generation of behaviour is modulated by factors influencing the internal state, as well as how responses are generated to events in the external environment.

Pathological changes can be the sole cause of behavioural change – indeed, behavioural signs such as lameness are common first indicators of disease in veterinary medicine. In some cases complex behavioural signs, such as aggressive behaviour towards an owner, can occur entirely as a consequence of pathological events, such as focal seizures. Although such events are rare, their characteristics need to be distinguished from behaviours generated in response to external stimuli when investigating the cause of undesired behaviours. However, physiological or pathological changes more commonly modify cats' behaviour rather than solely cause it. This is through alterations in one or more of the following: (i) perception of external events; (ii) the motivation to show a response; (iii) the threshold at which

© J.W.S. Bradshaw, R.A. Casey and S.L. Brown 2012. *The Behaviour of the Domestic Cat*, 2nd Edn (J.W.S. Bradshaw *et al.*)

a response is shown; and (iv) the manner in which a response is generated. Understanding the way that behaviour develops in individual cats, therefore, requires both an understanding of how behaviour is modulated through learning (see Chapter 3, this volume) and how disease can modify such processes. The physical examination of cases by a veterinary surgeon is therefore important before investigating individual cases of undesired behaviour (Fatjó and Bowen, 2009).

In addition to disease processes being important factors in the development of undesired behaviours, cats' responses to their environment can also influence the onset or development of disease. For example, stress can influence susceptibility to infectious disease, or the shedding of infectious agents from animals with carrier status (e.g. Addie *et al.*, 2009). Stress caused by environmental factors also influences immune functioning and the onset of bouts of chronic diseases such as feline idiopathic cystitis (Seawright *et al.*, 2008).

Since it is impractical to consider all possible medical causes of behavioural change here, we will use a functional approach to consider how disease processes may alter the motivation for behavioural responses in the cat, modify their occurrence or generate new responses. First, we consider how physiological or pathological changes may alter an individual cat's relative motivation to show behaviour, hence changing its frequency or timing. Next, we examine how normal behavioural responses to external events can be modified by internal factors, and where behaviours are generated by disease states. Finally, we give examples of how individual responses to the environment can influence the onset of disease in the cat.

Effects on Motivation to Show Normal Behaviour

Increased or decreased motivation to show normal responses, or the initiation of new behaviours, can all occur in association with physiological and pathological changes. Normal physiological variations that may influence the initiation of behaviour include the fluctuation of female reproductive hormones over the oestrous cycle in entire females. For example, profound changes in the behaviour of entire queens occur with the oestrous cycle, notably related to mate seeking (see Chapter 8, this volume), and this may or may not be seasonal depending on regional photoperiods (Faya *et al.*, 2011).

Disease processes can also modify the initiation of behaviour by altering underlying motivation. For example, a cat's motivation to acquire food may be increased by endocrine disorders such as hyperthyroidism (Salisbury, 1991). In such cases, behaviours aimed at obtaining food, such as vocalizing or rubbing around the owner, occur more frequently or in different spatial or temporal contexts. Similarly, conditions decreasing appetite may reduce food-soliciting behaviours towards the owner. Excessive eating or drinking caused by a medical condition will inevitably lead to a higher frequency of toileting behaviour. Hence a cat with a condition causing polydipsia (excessive drinking) may potentially show signs of inappropriate urination, because access to toileting sites becomes limiting with the increased need to eliminate. Similarly in conditions affecting metabolic rate or body temperature, cats may show altered motivation to seek sources of heat (e.g. in hyperthyroidism) or cool locations (e.g. fever). In some cases such changes can lead to the onset of undesired behaviours, perhaps in combination with situational factors. For example, a cat in a household with others that are not socially compatible may only be able to access a litter box intermittently, but concurrent disease-induced polyuria may bring about a change to an 'inappropriate' toileting site due to insufficient access for toileting needs.

Pain is one of the most common reasons for cats to alter their motivation to show behaviours. For example, cats may have a reduced desire to play or interact when suffering from joint pain, and this may be the first sign of which owners are aware in the development of osteoarthritis in older cats (Lascelles and Robertson, 2010). Pain may also lead to context-specific responses: for example, a cat with a bite abscess on the tail base may avoid being stroked by its owners, perhaps by moving away or even showing aggression to stop owner contact. Pain is processed through a network of structures known as the 'pain matrix' (Jones *et al.*, 2003). The system has two parallel pathways, one of which (the medial pathway) is involved in the emotional component (i.e. sensation) of pain; the other (lateral) pathway is involved in discriminating the particular details of the stimulus (i.e. type of sensation, location, intensity and duration). Both pathways involve the thalamus and pass through to the insular and somatosensory cortices, respectively (Kulkarni *et al.*, 2005). On contact with a painful stimulus, the areas of the brain associated with fear and anxiety are activated (Rainville, 2002); this is obviously adaptive as it enables the animal to learn the salient aspects of an environment that lead to a pain-eliciting outcome.

Altered motivation to groom areas of the body is also a common result of conditions causing irritation to the skin, and thus activation of mechanoreceptors (see Chapter 2, this volume). A range of conditions including parasite infestation and atopic, infectious or autoimmune-mediated skin disease can result in a cat showing increased grooming, biting or other skin-directed behaviour.

In cases of general illness or infection, animals commonly show what are termed 'sickness behaviours' (Johnson, 2002), which have a range of functions such as conserving energy or promoting defensive mechanisms in the body such as fever (Broom, 2006). Response to infection is mediated by the release of cytokines, which have a direct effect on the brain, leading to increased sleep and reduced activity. Cytokines are also suggested to influence activation of the hypothalamo–pituitary–adrenal axis and hence modulate behavioural responses (McCann *et al.*, 2000). Reduced motivation to be active or interact socially may also occur with sensory deficits, such as loss of sight (Turner, 2004).

Modulation of Existing Responses to External Events

In many cases, changes in internal state through physiological variation or disease do not lead to the initiation of a behaviour, but modify existing behavioural responses to external events. These effects can occur through the input of sensory information to the brain, the evaluation of information or in the generation of a motor response.

Modulation of sensory information inputting the brain

Information entering the brain is generated in the sensory organs such as the retinal cells in the eye, the nasal epithelium or the sensory receptors in the skin (see Chapter 2, this volume). Conditions that influence the relative activation of these organs by external stimuli, or the passage of impulses along the sensory nerves to the sensory cortices of the brain, will influence the extent to which cats respond to external events. For example, Siamese cats have an abnormal development of sensory input from the retina to the lateral geniculate nucleus, which means that their capacity to use stereoscopic vision to determine depth (see Chapter 2, this volume) is limited, and it is believed that the development of a squint in some individuals is to enable some overlap of visual fields (Hubel and Wiesel, 1971; Chapter 2, this volume). While the plasticity of the sensory cortex enables most Siamese to respond apparently normally to external events, reduced or absent binocular capability is likely to modify when and how individual cats respond to events, and potentially influence their hunting ability. Perception of olfactory information is dependent on the integrity of the nasal epithelium, and chronic damage from infectious upper respiratory tract disease can influence the relative perception of scents in the environment (Scherk, 2010), influencing a range of scent-related behaviours such as toileting and social interaction (see Chapters 5 and 8, this volume).

Modulation of sensory information from the integument is another example of where abnormalities can influence the extent to which cats perceive external stimuli. Changes in the relative activity of peripheral mechanoreceptors can result in either altered sensation (dysaesthesia) or exaggerated response to stimulation (hyperaesthesia) (Rizzo *et al.*, 1996). These changes in relative activation of sensory neurons arise as a result of changes in neuronal conduction properties with variation in sub-type of sodium ion channels in the nerve membrane (Waxman *et al.*, 2000). Damage to a peripheral sensory nerve may result, for example, in relative hyperpolarization, such that the threshold of activation by touch sensation is reduced (Matzner and Devor, 1992). These changes in peripheral nerve activation thresholds following damage are one potential underlying cause for behavioural changes in cats sometimes described as 'hyperaesthesia syndrome' or 'hyperkinesis' (Shell, 1994), where cats present with twitching or rippling skin and may jump around to groom intensely, as if responding to severe irritation, following only a mild touch sensation.

Factors influencing processing and evaluation of sensory information

Behaviour arising as a consequence of external changes can also be modulated by factors that influence how information is centrally processed or appraised. Normal physiological changes can influence these processes, such as fluctuations in reproductive hormones, since sex steroids act as modulatory neurotransmitters across many regions of the brain (Rupprecht and Holsboer, 1999). For example, evidence from other species suggests that changes in oestrogen may have profound effects on the activity of a range of different neurotransmitters at the cellular level, resulting in an increased chance of behavioural responses to external events and a reduced threshold of response to painful stimuli (Aloisi and Bonifazi, 2006).

Pathologies can also influence how sensory information is evaluated, and change either the threshold for the behaviour or the degree of behavioural response. Endocrine abnormalities can have a profound effect on responses to external events, as well as a range of physical signs. In Cushing's disease, for example, overproduction of glucocorticoids may ultimately reduce the production of corticotrophin-releasing factor in the hypothalamus due to negative feedback, resulting in an animal that is unresponsive to external events and lethargic. Corticosteroids are also commonly used in the treatment of various medical conditions in cats, and evidence from human patients suggests that iatrogenic steroids may also have effects on mood and behaviour (Brown and Chandler, 2001).

Diseases that influence the breakdown and metabolism of dietary components can also have an influence on the threshold of response generation in the brain. The most common examples are cases where either hepatic (liver) or renal (kidney) function is compromised, resulting in a reduced ability to remove by-products of digestion from the circulation and excrete these via the urine. The behavioural and neurological consequences of such diseases are commonly termed 'encephalopathies'. The effects on brain function can arise through the build-up of compounds that are neurotoxic, or alternatively can be caused by the relative unavailability of amino acids necessary for neurotransmitter turnover. For example, increased ammonium salts in the circulation as a result of hepatic insufficiency cannot be metabolized to urea in the central nervous system, and are instead converted to glutamine, which ultimately leads to an excess of this amino acid in the brain (Albrecht *et al.*, 2007). Since glutamine is the precursor for the excitatory neurotransmitter glutamate, this may explain transient decreases in the threshold of responding (for example, increased aggression) associated with encephalopathies. In addition, chronic low-grade hyperammonaemia has been associated with memory deficits and difficulty in adapting to new environments in both humans and rats (Apelqvist *et al.*, 1999). Because increased levels of ammonia, and hence increased glutamine and glutamate, occur as a protein meal is digested and the by-products circulate in the bloodstream, behavioural signs associated with hyperammonaemia will often occur temporarily, associated with feeding. Ultimately, increasing levels of glutamine result in cerebral oedema (Albrecht *et al.*, 2007) and more obvious neurological symptoms, such as loss of motor coordination, depression, hysteria, pacing, circling, seizures and, ultimately, coma and death.

Alterations of normal behaviour can arise as a consequence of disorders of the central nervous system. The specific effects of lesions on behaviour will relate to the specific area of the central nervous system affected. For example, lesions affecting the pineal gland may influence the sleep–wake cycle (Uz *et al.*, 2003). In older cats, behavioural changes in response to external events are commonly caused by a range of clinical conditions including osteoarthritis, systemic hypertension (often secondary to chronic kidney disease or hyperthyroidism), hyperthyroidism or sensory deficits (Gunn-Moore, 2011). In some cases such changes may result from reduced motivation to show new behaviours, for example due to pain, as discussed earlier. There is also a tendency across species for cognitive ability to decline with increasing age. For example, in both humans (Cherry and Park, 1993) and dogs (Christie *et al.*, 2005) the ability to perform allocentric tasks (i.e. requiring reference to external landmarks) declines with age. However, in a proportion of cats more profound changes in cognitive ability and behaviour are associated with specific pathological changes analogous to Alzheimer's disease in people, known as 'cognitive dysfunction' (Landsberg *et al.*, 2010). Associated alterations in response to external events might include reduced social interaction with owners or, alternatively, increased dependency on owner attention and presence, loss of previously learnt associations, altered or 'mixed-up' responses to previously learnt cues (e.g. change in toileting substrate preference) or general disorientation.

Influences on motor outputs and initiation of behaviour

Disease processes may also modify behavioural responses to external events by altering the output of motor information from the brain and the activation of peripheral muscles via neuromuscular junctions. Changes in this element of the pathway may influence the cat's ability to show a desired behaviour, or alter the form of the behaviour. A cat's ability to show a response may be influenced by diseases affecting the motor cortex of the brain or spinal cord. For example, spinal damage can influence the motor control of elimination, potentially resulting in urination or defecation in undesired locations. Control of toileting can also be influenced by conditions affecting the gastrointestinal or urinary tract. Diseases causing inflammation in the bowel, or those affecting absorption, will potentially influence both the frequency and urgency of defecation.

Feline idiopathic cystitis (FIC), also known as idiopathic feline lower urinary tract disease (iFLUTD), is the most common medical cause of abnormal urination in the cat, and hence is an important differential diagnosis to consider when investigating cats presenting with inappropriate elimination (Buffington *et al.*,

1997). FIC is termed 'idiopathic' when there is no obvious physical cause to account for the condition, and is diagnosed by excluding other causes of lower urinary tract inflammation (such as urinary tract infection, urethral strictures, neoplasia or urolithiasis) (Kalkstein *et al.*, 1999). Inappropriate elimination may be the first presenting sign of FIC, and is thought to occur because the cat associates the pain of urination with the specific location in which it has previously urinated. In addition, the condition causes increased frequency and urgency of urination. Because the signs of the condition are commonly present during bouts of 3 or 4 days, affected cats may show repeated changes in the location of urination for several days in a row. Pain on urination can also cause the cat to appear distressed and vocalize before and during urination, and abdominal pain may make cats reluctant to be handled, either moving away or showing aggression as an avoidance response. Male cats may also change their posture from a squat to standing up, as squatting, which bends the urethra, may cause further discomfort (Seawright *et al.*, 2008).

The ability of cats to show desired behavioural responses can also be influenced by diseases affecting movement, such as degenerative joint disease (e.g. osteoarthritis), neuropathies of motor nerves (e.g. diabetic polyneuropathy), disorders of the neuromuscular junction (e.g. myasthenia gravis) or problems with muscular functions (e.g. myopathies associated with feline leukaemia virus).

Behaviours Occurring Entirely as a Result of Disease Processes

In some cases behavioural signs in cats are generated entirely as a result of a disease process, and occur unrelated to external events (Reisner, 1991). In general these types of behaviour are less common than where pathology acts to modify behavioural responses to external events. Abnormalities that can generate a response may occur in any part of the process by which a normal response is generated (i.e. sensory input to the brain, the processing of information in the central nervous system or the output of information from the motor cortex).

The generation of abnormal sensory information will create similar behavioural responses to sensory inputs based on actual external stimuli. For example, in addition to dysaesthesic and hyperaesthesic responses in peripheral sensory nerves (described previously), paraesthesic responses can occur, whereby spontaneous activity is generated in the nerve without any mechanical stimulation of the corresponding receptors. This abnormal or ectopic impulse generation is also commonly called sensory neuropathy. Because these abnormal inputs are processed as if they arise in response to normal external events, the reaction of the cat is appropriate to irritation or pain arising from the source (i.e. grooming, scratching or attempting to bite at the affected part of the body). In some cases these responses can be extreme and cause both cat and owner considerable distress. The tail is a common target for such behaviour, possibly because the sensory nerves in the tail are more susceptible to damage (e.g. through bite injuries or the tail being caught in a doorway). A cat may repeatedly and persistently attack its tail to the extent that it becomes damaged and requires repeated medical or surgical intervention. Depending on the origin of the sensory damage, and the extent to which sensitization of the response has occurred, the behaviour can even continue following amputation of part or all of the tail. An abnormal pain sensation in the face and mouth region has also been described, particularly in oriental breeds. Although there may be multiple possible factors involved in the development of this condition, damage to the facial nerves, such as the trigeminal nerve, is likely to be an important factor. Cats suffering from this condition show signs varying in severity; in the most extreme cases cats can cause considerable self-damage by clawing at their own face in an apparent attempt to alleviate the pain (Rusbridge *et al.*, 2010).

Spontaneous activity through ectopic activity in nerve cells also occurs in the brain. Spontaneous depolarization shifts or repetitive discharge from hyperexcitable collections of neurons in the brain are expressed as epileptic seizures. In the typical 'grand mal' seizure, this activity spreads across the whole brain. However, the spread of electrical activity can also be localized, influencing only one part of the brain. The behavioural response arising from these focal seizures will depend on the part of the brain that is affected and the extent to which it spreads (Dahl, 1999). Hence, a focal seizure in the visual cortex is likely to result in a cat responding as if it has seen something (i.e. a 'visual hallucination'). The response will often be identical to that as would occur if the animal had seen a 'real' stimulus, since activation of this brain region will lead to the generation of the same behavioural response as if the activation was due to inputs via the optic nerve. Hence, cats may appear to pounce on or chase objects which are not there. Seizure activity localized to the limbic part of the brain, responsible for the generation of emotional responses, can result in sudden unprovoked behavioural responses indicative of extreme emotional disturbance. For example, cats may suddenly run off and hide, or show extreme aggression to the nearest person, object or other animal. Where the focus of focal seizures is the motor cortex, resulting behaviours are often rather more fixed in form, rather than the more 'goal-oriented' and variable behaviours associated with sensory focal seizures.

Seizure activity develops over time through the process of 'kindling', whereby spontaneous activity in one area gradually results in the 'recruitment' of neighbouring cells (Bertram, 2007). In the early stages, seizure activity may therefore result in relatively mild signs which are often disregarded. For example, humans with seizures in the limbic part of the brain may initially report mild 'feelings' occurring spontaneously, which develop over time into more obvious seizures with clear behavioural indicators (Bertram, 2007). The plasticity of the brain, and susceptibility of cells to repeated electrical stimulation, means that behavioural changes may be used to identify early seizure activity (Shihab *et al.*, 2011). The other factor to consider with epilepsy is the effect of seizures on behaviour occurring between the episodes. Because the action of the seizures themselves is to partially 'remodel' areas of the brain, the response of the animal to normal stimuli in the interictal (between seizure) period can also be affected (Adamec, 2003).

Persistence of Behavioural Outcomes of Disease

When evaluating the behaviour of an individual animal, it is important not only to consider currently occurring physiological or pathological impacts on behaviour, but to investigate the potential impact of historical effects. In many cases, behavioural changes can arise as a consequence of a previous condition that has been resolved, but where the behavioural response has been retained through learning. For example, male cats with FIC may stand to urinate in order to facilitate the passage of inflammatory material, or due to urethral spasm. However, where these individuals learn that this posture eases the pain associated with elimination, they may continue to show this posture even after resolution of the disease. Similarly, avoidance responses learnt when a cat was in pain may be retained after resolution of the painful lesion. Where aggression to an owner stroking the caudal dorsum effectively prevents contact on the tail base, a cat may continue to respond aggressively after resolution of a tail base abscess because it has not had the opportunity to learn that contact in this area is no longer painful. Behaviours originating for medical reasons can also become reinforced and established for other reasons. For example, grooming behaviour initiated because of an irritating lesion could become reinforced by the owner paying the cat attention when the behaviour is shown. This type of reinforcement is particularly likely where cats highly value owner attention, as response from the owner will be a valued outcome.

Environmental Stressors in Disease Aetiology

It is increasingly recognized in human medicine that psychological stressors impact extensively on somatic disease (Nater *et al.*, 2006). Similarly, in veterinary medicine, the influence on disease processes of environment, and individual responses to environmental change, have gained in recognition (Casey, 2010). However, much still needs to be investigated to elucidate the relationship between stress exposure and disease susceptibility. It is unclear, for example, why some individuals appear to be more susceptible to somatic disease than others exposed to similar environments. In addition, it is not known why different individuals exposed to the same stressors develop conditions involving different body systems. However, it is likely that some individuals are predisposed to vulnerability, such that co-occurrence with extreme environments leads to dysregulated stress responses, resulting in disease. Further research is needed to investigate the nature of this vulnerability, and how it interacts with stressors that lead to such consequences.

In cats, FIC is the condition that is most widely recognized as associated with exposure to stress (Seawright *et al.*, 2008). Indeed, differentiating cases of FIC from inappropriate elimination can be complicated, as exposure to events that cats find aversive is an important 'flare factor' in the multifactorial aetiology of this condition (Cameron *et al.*, 2004), and hence similar factors can lead to both inappropriate elimination and FIC. Other chronic conditions in which stress exposure is considered a factor in humans include irritable bowel syndrome (Murray *et al.*, 2004) and chronic skin disease (Kimyai-Asadi and Usman, 2001), both of which are anecdotally also relevant in cats. The hyperaesthetic sensory responses, as described earlier, are also postulated to have environmental stress as an aetiological factor.

In addition to chronic disease conditions, where clinical signs appear to be exacerbated or precipitated by stressful events, acute stress can have important impacts on susceptibility to infectious disease. For example, the high level of stress experienced by many cats when entering a boarding or rescue cattery, associated with loss of predictability and change in environmental cues, leads to an increased susceptibility to infectious disease, at the same time as they are brought into the proximity of other cats, thus increasing the probability of contact with infectious agents.

Conclusions

Important welfare implications arise from the appropriate recognition of the influences of disease on behaviour. Behavioural signs may be the first indicators of disease and enable the early treatment of potentially serious medical conditions. In addition, the recognition of confounding disease processes will ensure that behavioural interventions are considered that take these into account, and better inform prognosis when treating undesired behaviours. Furthermore, where stress is likely to impact on disease risk, interventions to reduce exposure to stressors can have important health impacts.

In considering the relationship between pathological disease or physiological variation and behavioural signs it is important to consider not only those behaviours that are directly caused by pathologies, but also those where changes in internal state may modify existing responses. In addition, it is imperative to adopt a rational, functional approach and consider those aspects of normal response generation where pathology could be having an effect. For example, where a cat is observed attacking its tail, the important factors to investigate and differentiate from behavioural causes will range from conditions that cause irritation of the skin right through to those that might be generating an abnormal motor response.

References

Abrahams, V.C., Hodgkins, M. and Downey, D. (1987) Morphology, distribution and density of sensory receptors in the glabrous skin of the cat rhinarium. *Journal of Morphology* 191, 109–114.

Adamec, R. (2003) Kindling induced lasting interictal alterations of affective behavior. *Annals of the New York Academy of Sciences* 985, 495–497.

Adamec, R.E. (1976) The interaction of hunger and preying in the domestic cat (*Felis catus*): an adaptive hierarchy? *Behavioural Biology* 18, 263–272.

Adamec, R.E., Stark-Adamec, C. and Livingstone, K.E. (1980) The development of predatory aggression and defense in the domestic cat (*Felis catus*). *Behavioural and Neural Biology* 30, 389–409.

Adamec, R.E., Stark-Adamec, C. and Livingstone, K.E. (1983) The expression of an early developmentally emergent defensive bias in the adult domestic cat (*Felis catus*) in non-predatory situations. *Applied Animal Ethology* 10, 89–108.

Adamelli, S., Marinelli, L., Normando, S. and Bono, G. (2005) Owner and cat features influence the quality of life of the cat. *Applied Animal Behaviour Science* 94, 89–98.

Addie, D., Belák, S., Bourcraut-Baralon, C., Egberin, H., Frymus, T., Gruffydd-Jones, T. *et al.* (2009) Feline infectious peritonitis: ABCD guidelines on prevention and management. *Journal of Feline Medicine and Surgery* 11, 594–604.

Albrecht, J., Sonnewald, U., Waagepetersen, H.S. and Schousboe, A. (2007) Glutamine in the central nervous system: function and dysfunction. *Frontiers in Bioscience* 12, 332–343.

Aldis, O. (1975) *Play Fighting*. Academic Press, New York.

Aloisi, A.M. and Bonifazi, M. (2006) Sex hormones, central nervous system and pain. *Hormones and Behaviour* 50, 1–7.

APBC (2005) Annual review of cases 2005. Available at: http://www.apbc.org.uk/sites/default/files/review_2005.pdf (accessed 11 September 2012).

Apelqvist, G., Hindfelt, B., Andersson, G. and Bengtsson, F. (1999) Altered adaptive behaviour expressed in an open-field paradigm in experimental hepatic encephalopathy. *Behavioural Brain Research* 106, 165–173.

Appleby, M.C. (1993) How animals perceive a hierarchy: reactions to Freeman *et al.* *Animal Behaviour* 46, 1232–1233.

Archer, J. (2011) Pet keeping: a case study in maladaptive behaviour. In: Salmon, C.A. and Shackelford, T.K. (eds) *The Oxford Handbook of Evolutionary Family Psychology.* Oxford University Press, New York, pp. 281–296.

Bacon, B.A., Lepore, F. and Guillemot, J.-P. (1999) Binocular interactions and spatial disparity sensitivity in the superior colliculus of the Siamese cat. *Experimental Brain Research* 124, 181–192.

Bang, A., Deshpande, S., Sumana, A. and Gadagkar, R. (2010) Choosing an appropriate index to construct dominance hierarchies in animal societies: a comparison of three indices. *Animal Behaviour* 79, 631–636.

Bard, P. and Macht, M.B. (1958) The behaviour of chronically decerebrate cats. In: Wolstenholme, G.E.W. and O'Connor, C.M. (eds) *CIBA Foundation Symposium on the Neurological Basis of Behaviour.* J. & A. Churchill, Ltd, London, pp. 55–71.

Barratt, D. (1997) Home range size, habitat utilization and movement patterns of suburban and farm cats (*Felis catus*). *Ecography* 20, 271–280.

Barrett, P. and Bateson, P. (1978) The development of play in cats. *Behaviour* 66, 106–120.

Barry, K. and Crowell-Davis, S. (1999) Gender differences in the social behavior of the neutered indoor-only domestic cat. *Applied Animal Behaviour Science* 64, 193–211.

Bateson, P. (1979) How do sensitive periods arise and what are they for? *Animal Behaviour* 27, 470–486.

Bateson, P. (2000) Behavioural development in the cat. In: Turner, D.C. and Bateson, P. (eds) *The Domestic Cat: The Biology of its Behaviour,* 2nd Edn. Cambridge University Press, Cambridge, pp. 9–22.

Bateson, P. and Bateson, M. (2002) Post-weaning feeding problems in young domestic cats – a new hypothesis. *The Veterinary Journal* 163, 113–114.

Bateson, P. and Turner, D.C. (1988) Questions about cats. In: Turner, D.C. and Bateson, P. (eds) *The Domestic Cat: The Biology of its Behaviour.* Cambridge University Press, Cambridge, pp. 193–201.

Bateson, P., Mendl, M. and Feaver, J. (1990) Play in the domestic cat is enhanced by rationing of the mother during lactation. *Animal Behaviour* 40, 514–525.

Belkin, M., Yinon, U., Rose, L. and Reisert, I. (1977) Effect of visual environment on refractive error of cats. *Documenta Ophthalmologica* 42, 433–437.

Berkley, M.A. (1976) Cat visual psychophysics: neural correlates and comparisons with man. *Progress in Psychobiology and Physiological Psychology* 6, 63–119.

Bernstein, I.L. (1999) Taste aversion learning: a contemporary perspective. *Nutrition* 15, 229–234.

Bernstein, P.L. (2007) The human–cat relationship. In: Rochlitz, I. (ed) *The Welfare of Cats.* Springer Press, Dordrecht, The Netherlands, pp. 47–89.

Bernstein, P.L. and Strack, M. (1996) A game of cat and house: spatial patterns and behavior of 14 domestic cats (*Felis catus*) in the home. *Anthrozoos* 9, 25–39.

Bertram, E. (2007) The relevance of kindling for human epilepsy. *Epilepsia* 48, 65–74.

Biben, M. (1979) Predation and predatory play behaviour of domestic cats. *Animal Behaviour* 27, 81–94.

Blakemore, C. and Van Sluyters, R.C. (1975) Innate and environmental factors in the development of the kitten's visual cortex. *Journal of Physiology* 248, 663–716.

Bonnaud, E., Medina, F.M., Vidal, E., Nogales, M., Tershy, B., Zavaleta, E.S. *et al.* (2011) The diet of feral cats on islands: a review and a call for more studies. *Biological Invasions* 13, 581–603.

Borchelt, P.L. and Voith, V.L. (1982) Classification of animal behaviour problems. *Veterinary Clinics of North America: Small Animal Practice* 12, 571–585.

Boudreau, J.C. (1989) Neurophysiology and stimulus chemistry of mammalian taste systems. In: Teranishi, R., Buttery, R.G. and Shahidi, F. (eds) *Flavor Chemistry: Trends and Developments*, ACS Symposium Series 388, pp. 122–137.

Bradshaw, J.W.S. (1986) Mere exposure reduces cats' neophobia to unfamiliar food. *Animal Behaviour* 34, 613–614.

Bradshaw, J.W.S. (1991) Sensory and experiential factors in the design of foods for domestic dogs and cats. *Proceedings of the Nutrition Society* 50, 99–106.

Bradshaw, J.W.S. (2006) The evolutionary basis for the feeding behavior of domestic dogs (*Canis familiaris*) and cats (*Felis catus*). *Journal of Nutrition* 136, 1927S–1931S.

Bradshaw, J.W.S. and Cameron-Beaumont, C. (2000) The signalling repertoire of the domestic cat and its undomesticated relatives. In: Turner, D.C. and Bateson, P. (eds) *The Domestic Cat: The Biology of its Behaviour*, 2nd Edn. Cambridge University Press, Cambridge, pp. 68–93.

Bradshaw, J.W.S. and Casey, R.A. (2007) Anthropomorphism and anthropocentrism as influences in the quality of life of companion animals. *Animal Welfare* 16(S), 149–154.

Bradshaw, J.W.S. and Cook, S. (1996) Patterns of pet cat behaviour at feeding occasions. *Applied Animal Behavior Science* 47, 61–74.

Bradshaw, J.W.S. and Hall, S.L. (1999) Affiliative behaviour of related and unrelated pairs of cats in catteries: a preliminary report. *Applied Animal Behaviour Science* 63, 251–255.

Bradshaw, J.W.S. and Limond, J. (1997) Attachment to cats and its relationship with emotional support: a cross cultural study. In: *Proceedings of the ISAZ Conference*, Boston, Massachusetts, 24–25 July.

Bradshaw, J.W.S. and Lovett, R.E. (2003) Dominance hierarchies in domestic cats: useful construct or bad habit? *Proceedings of the British Society of Animal Science Conference 2003*, 16.

Bradshaw, J.W.S., Horsfield, G.F., Allen, J.A. and Robinson, I.H. (1999) Feral cats: their role in the population dynamics of *Felis catus*. *Applied Animal Behaviour Science* 65, 273–283.

Bradshaw, J.W.S., Casey, R.A. and MacDonald, J.M. (2000a) The occurrence of unwanted behaviour in the pet cat population. *Proceedings of the Companion Animal Behaviour Therapy Study Group Study Day*, Birmingham, UK, pp. 41–42.

Bradshaw, J.W.S., Healey, L.M., Thorne, C.J., Macdonald, D.W. and Arden-Clark, C. (2000b) Differences in food preferences between individuals and populations of domestic cats *Felis silvestris catus*. *Applied Animal Behaviour Science* 68, 257–268.

Bradshaw, J.W.S., Blackwell, E.J. and Casey, R.A. (2009) Dominance in domestic dogs – useful construct or bad habit? *Journal of Veterinary Behavior* 4, 135–144.

Bravo, M., Blake, R. and Morrison, S. (1988) Cats see subjective contours. *Vision Research* 18, 861–865.

Broom, D.M. (1988) The scientific assessment of animal welfare. *Applied Animal Behaviour Science* 20, 5–19.

Broom, D.M. (2006) Behaviour and welfare in relation to pathology. *Applied Animal Behaviour Science* 97, 73–83.

Brown, E.S. and Chandler, P.A. (2001) Mood and cognitive changes during systemic corticosteroid therapy. *Primary Care Companion. Journal of Clinical Psychiatry* 3, 17–21.

Brown, K.A., Buchwald, J.S., Johnson, J.R. and Mikolich, D.J. (1978) Vocalization in the cat and kitten. *Developmental Psychobiology* 11, 559–570.

Brown, S.L. (1993) The social behaviour of neutered domestic cats (*Felis catus*). PhD thesis, University of Southampton, Southampton, UK.

Brown, S.L. and Bradshaw, J.W.S. (1996) Social behaviour in a small colony of feral cats. *Journal of the Feline Advisory Bureau* 34, 35–37.

Buesching, C.D., Stopka, P. and Macdonald, D.W. (2003) The social function of allomarking in the European badger (*Meles meles*). *Behaviour* 140, 965–980.

Buffington, C.A., Chew, D.J., Kendall, M.S., Scrivani, P.V., Thompson, S.B., Blaisdell, J.L. *et al.* (1997) Clinical evaluation of cats with nonobstructive urinary tract diseases. *Journal of American Veterinary Medical Association* 210, 46–50.

Burgess, P.R. and Perl, E.R. (1973) Cutaneous mechanoreceptors and nociceptors. In: Iggo, A. (ed.) *Handbook of Sensory Physiology*, Vol. II, The Somatosensory System. Springer-Verlag, New York, pp. 29–78.

Cafazzo, S. and Natoli, E. (2009) The social function of tail up in the domestic cat (*Felis sylvestris catus*). *Behavioural Processes* 80, 60–66.

Cameron, M.E., Casey, R.A., Bradshaw, J.W.S., Waran, N.K. and Gunn-Moore, D.A. (2004) A study of environmental and behavioural factors that may be associated with feline idiopathic cystitis. *Journal of Small Animal Practice* 45, 144–147.

Cameron-Beaumont, C.L. (1997) Visual and tactile communication in the domestic cat (*Felis silvestris catus*) and undomesticated small felids. PhD thesis, University of Southampton, Southampton, UK.

Cameron-Beaumont, C., Lowe, S.E. and Bradshaw, J.W.S. (2002) Evidence suggesting preadaptation to domestication throughout the small Felidae. *Biological Journal of the Linnean Society* 75, 361–366.

Caro, T.M. (1980) Effects of the mother, object play, and adult experience on predation in cats. *Behavioural and Neural Biology* 29, 29–51.

Caro, T.M. (1981) Predatory behaviour and social play in kittens. *Behaviour* 76, 1–24.

Caro, T.M. (1989) Determinants of asociality in felids. In: Standen, V. and Foley, R.A. (eds) *Comparative Socioecology: the Behavioural Ecology of Humans and Other Mammals*. Blackwell Scientific Publications, Oxford, UK, pp. 41–74.

Carpenter, J.A. (1956) Species differences in taste preferences. *Journal of Comparative and Physiological Psychology* 49, 139–144.

Casey, R. (2008) Undesirable behaviours in the domestic cat: development, consequences and treatment. PhD thesis, University of Bristol, Bristol, UK.

Casey, R.A. (2010) Fear, anxiety and conflict in companion animals. In: Lindley, S. and Watson, P. (eds) *BSAVA Manual of Canine and Feline Rehabilitation, Supportive and Palliative Care*. British Small Animal Veterinary Association, Quedgeley, UK, pp. 31–41.

Casey, R.A. and Bradshaw, J.W.S. (2001) A comparison of referred feline clinical behaviour cases with general population prevalence data. *Scientific Proceedings of the British Small Animal Veterinary Association Congress*, Birmingham, UK, p. 529.

Casey, R.A. and Bradshaw, J.W.S. (2005) The assessment of welfare. In: Rochlitz, I. (ed.) *The Welfare of Cats*. Springer, Dordrecht, The Netherlands, pp. 23–46.

Casey, R.A. and Bradshaw, J.W.S. (2008) The effects of additional socialisation for kittens in a rescue centre on their behaviour and suitability as a pet. *Applied Animal Behaviour Science* 114, 196–205.

Casey, R.A., Vandenbussche, S., Bradshaw, J.W.S. and Roberts, M.A. (2009) Reasons for relinquishment and return of domestic cats (*Felis silvestris catus*) to rescue shelters in the UK. *Anthrozoos* 22, 347–358.

Chaadaeva, E.V. and Sokolova, N.N. (2005) Development of vocal repertoire in kittens of *Felis lybica* and *F.catus*. *Zoologicheskii Zhu* 84, 1402–1415.

Cherry, K.E. and Park, D.C. (1993) Individual difference and contextual variables influence spatial memory in younger and older adults. *Psychology and Aging* 8, 517–526.

Chesler, P. (1969) Maternal influence in learning by observation in kittens. *Science* 166, 901–903.

Christie, L.-A., Studzinski, C.M., Araujo, J.A., Leung, C.S.K., Ikeda-Douglas, C.J., Head, E. *et al.* (2005) A comparison of egocentric and allocentric age-dependent spatial learning in the beagle dog. *Progress in Neuro-Psychopharmacology and Biological Psychiatry* 29, 361–369.

Church, S.C., Allen, J.A. and Bradshaw, J.W.S. (1996) Frequency-dependent food selection by domestic cats: a comparative study. *Ethology* 102, 495–509.

Clark, C.C.A., Gruffydd-Jones, T. and Murray, J.K. (2012) Number of cats and dogs in UK welfare organisations. *Veterinary Record* 170, 493.

Clark, J.M. (1975) The effects of selection and human preference on coat colour gene frequencies in urban cats. *Heredity* 35, 195–210.

Clifford, C.W.G. and Ibbotson, M.R. (2003) Fundamental mechanisms of visual motion detection: models, cells and functions. *Progress in Neurobiology* 68, 409–437.

Clutton-Brock, J. (1987) *A Natural History of Domesticated Mammals.* Cambridge University Press, Cambridge, and the British Museum (Natural History), London.

Colgan, P. (1989) *Animal Motivation.* Chapman & Hall, London.

Collard, R.R. (1967) Fear of strangers and play behavior in kittens with varied social experience. *Child Development* 38, 877–891.

Collier, G., Johnson, D.F. and Morgan, C. (1997) Meal patterns of cats encountering variable food procurement cost. *Journal of the Experimental Analysis of Behavior* 67, 303–310.

Cooper, L.L. (1997) Feline inappropriate elimination. *Veterinary Clinics of North America: Small Animal Practice* 27, 569–600.

Costalupes, J.A. (1983) Temporal integration of pure tones in the cat. *Hearing Research* 9, 43–54.

Crouch, J.E. (1969) *Text Atlas of Cat Anatomy.* Lea and Febiger, Philadelphia, Pennsylvania.

Crowell-Davis, S.L. Barry, K. and Wolfe, R. (1997) Social behaviour and aggressive problems of cats. *Veterinary Clinics of North America: Small Animal Practice* 27, 549–568.

Crowell-Davis, S.L., Curtis, T. and Knowles, R. (2004) Social organization in the cat: a modern understanding. *Journal of Feline Medicine and Surgery* 6, 19–28.

Curtis, T., Knowles, R. and Crowell-Davis, S. (2003) Influence of familiarity and relatedness on proximity and allogrooming in domestic cats (*Felis catus*). *American Journal of Veterinary Research* 64, 1151–1154.

Dahl, J.C. (1999) A behaviour medicine approach to epilepsy – time for a paradigm shift? *Scandinavian Journal of Behaviour Therapy* 28, 97–114.

Dantas-Divers, L.M.S., Crowell-Davis, S.L., Alford, K., Genaro, G., D'Almeida, J.M. and Paixao, R.L. (2011) Agonistic behavior and environmental enrichment of cats communally housed in a shelter. *Journal of the American Veterinary Medical Association* 239, 796–802.

Dards, J.L. (1983) The behaviour of dockyard cats: interactions of adult males. *Applied Animal Ethology* 10, 133–153.

Davey, G. (1989) *Ecological Learning Theory.* Routledge, London.

de Boer, J.N. (1977a) Dominance relations in pairs of domestic cats. *Behavioural Processes* 2, 227–242.

de Boer, J.N. (1977b) The age of olfactory cues functioning in chemocommunication among male domestic cats. *Behavioural Processes* 2, 209–225.

Deag, J.M., Manning, A. and Lawrence, C.E. (1988) Factors influencing the mother–kitten relationship. In: Turner, D.C. and Bateson, P. (eds) *The Domestic Cat: The Biology of its Behaviour.* Cambridge University Press, Cambridge, pp. 23–39.

DeAngelis, G.C. (2000) Seeing in three dimensions: the neurophysiology of stereopsis. *Trends in Cognitive Sciences* 4, 80–90.

Deliagina, T.G., Orlovsky, G.N., Zelenin, P.V. and Beloozerova, I.N. (2006) Neural bases of postural control. *Physiology* 21, 216–225.

Devillard, S., Say, L. and Pontier, D. (2003) Dispersal pattern of domestic cats (*Felis catus*) in a promiscuous urban population: do females disperse or die? *Journal of Animal Ecology* 72, 203–211.

Dickman, C.R. (2009) House cats as predators in the Australian environment: impacts and management. *Human–Wildlife Conflicts* 3, 41–48.

Drews, C. (1993) The concept and definition of dominance in animal behaviour. *Behaviour* 125, 283–313.

Driscoll, C.A., Menotti-Raymond, M., Roca, A.L., Hupe, K., Johnson, W.E., Geffen, E. *et al.* (2007) The Near Eastern origin of cat domestication. *Science* 317, 519–523.

Dukas, R. (2002) Behavioural and ecological consequences of limited attention. *Philosophical Transactions of the Royal Society of London B* 357, 1539–1547.

Dumas, C. (1992) Object permanence in cats (*Felis catus*): an ecological approach to the study of invisible displacements. *Journal of Comparative Psychology* 106, 404–410.

Dumas, C. (2000) Flexible search behavior in domestic cats (*Felis catus*): a case study of predator–prey interaction. *Journal of Comparative Psychology* 114, 232–238.

Dumas, C. and Dore, F.Y. (1991) Cognitive development in kittens (*Felis catus*): an observational study of object permanence and sensorimotor intelligence. *Journal of Comparative Psychology* 105, 357–365.

Durr, R. and Smith, C. (1997) Individual differences and their relation to social structure in domestic cats. *Journal of Comparative Psychology* 111, 412–418.

Dybdall, K., Strasser, R. and Katz, T. (2007) Behavioral differences between owner surrender and stray domestic cats after entering an animal shelter. *Applied Animal Behaviour Science* 104, 85–94.

Eckstein, R.A. and Hart, B.L. (2000) The organisation and control of grooming in cats. *Applied Animal Behaviour Science* 68, 131–140.

Edney, A.T.B. (ed.) (1988) *The Waltham Book of Dog and Cat Nutrition*, 2nd Edn. Pergamon Press, Oxford, UK.

Eisert, R. (2011) Hypercarnivory and the brain: protein requirements of cats reconsidered. *Journal of Comparative Physiology B* 181, 1–17.

Ellis, S. (2009) Environmental enrichment: practical strategies for improving animal welfare. *Journal of Feline Medicine and Surgery* 11, 901–912.

Elul, R. and Marchiafava, P.L. (1964) Accommodation of the eye as related to behaviour in the cat. *Archives Italienne de Biologie* 102, 616–644.

Everett, G.M. (1944) Observations on the behavior and neurophysiology of acute thiamin deficient cats. *American Journal of Physiology* 141, 439–448.

Evinger, C. and Fuchs, A.F. (1978) Saccadic, smooth pursuit and optokinetic eye movements of the trained cat. *Journal of Physiology* 285, 209–229.

Ewer, R.F. (1973) *The Carnivores.* Weidenfeld and Nicolson, London.

Fatjó, J. and Bowen, J. (2009) Medical and metabolic influences on behavioural disorders. In: Horwitz, D.F. and Mills, D.S. (eds) *BSAVA Manual of Canine and Feline Behavioural Medicine*, 2nd Edn. British Small Animal Veterinary Association, Quedgeley, UK, pp. 1–9.

Faure, E. and Kitchener, A.C. (2009) An archaeological and historical review of the relationship between felids and people. *Anthrozoös* 22, 221–238.

Fay, R.R. (1988) Comparative psychoacoustics. *Hearing Research* 34, 295–306.

Faya, M., Carranza, A., Priotto, M., Abeya, M., Diaz, J.D. and Gobello, C. (2011) Domestic queens under natural temperate photoperiod do not manifest seasonal anestrus. *Animal Reproduction Science* 129, 78–81.

Feaver, J., Mendl, M. and Bateson, P. (1986) A method for rating the individual distinctiveness of domestic cats. *Animal Behaviour* 34, 1016–1025.

Feldman, H. (1993) Maternal care and differences in the use of nests in the domestic cat. *Animal Behaviour* 45, 13–23.

Feldman, H. (1994a) Domestic cats and passive submission. *Animal Behaviour* 47, 457–459.

Feldman, H. (1994b) Methods of scent marking in the domestic cat. *Canadian Journal of Zoology* 72, 1093–1099.

Finkler, H. and Terkel, J. (2010) Cortisol levels and aggression in neutered and intact free-roaming female cats living in urban social groups. *Physiology and Behavior* 99, 343–347.

Finkler, H., Gunther, I. and Terkel, J. (2011) Behavioral differences between urban feeding groups of neutered and sexually intact free-roaming cats following a trap–neuter–return procedure. *Journal of the American Veterinary Medical Association* 238, 1141–1148.

Firestein, S. (2001) How the olfactory system makes sense of scents. *Nature* 413, 211–218.

Fiset, S. and Doré, F.Y. (1996) Spatial encoding in domestic cats (*Felis catus*). *Journal of Experimental Psychology: Animal Behavior Processes* 22, 420–437.

Fiset, S. and Doré, F.Y. (2006) Duration of cats' (*Felis catus*) working memory for disappearing objects. *Animal Cognition* 9, 62–70.

Fowler, G.A. and Sherk, H. (2003) Gaze during visually-guided locomotion in cats. *Behavioural Brain Research* 139, 83–96.

Frank, M.C. (2011) Sleep and developmental plasticity are not just for kids. *Progress in Brain Research* 193, 221–232.

Fraser, D. (2009) Assessing animal welfare: different philosophies, different scientific approaches. *Zoo Biology* 28, 507–518.

Frazer Sissom, D.E., Rice, D.A. and Peters, G. (1991) How cats purr. *Journal of Zoology*, London 223, 67–78.

Gálvez-López, E., Maes, L.D. and Abourachid, A. (2011) The search for stability on narrow supports: an experimental study in cats and dogs. *Zoology* 114, 224–232.

Geigy, C.A., Heid, S., Steffen, F., Danielson, K., Jaggy, A. and Gaillard, C. (2007) Does a pleiotropic gene explain deafness and blue irises in white cats? *Veterinary Journal* 173, 548–553.

German, A.J. (2006) The growing problem of obesity in dogs and cats. *Journal of Nutrition* 136, 1940S–1946S.

Gittleman, J.L. (1991) Carnivore olfactory bulb size: allometry, phylogeny and ecology. *Journal of Zoology, London* 225, 253–272.

Gordon, G. and Jukes, M.G.M. (1964) Dual organisation of the exteroceptive components of the cat's gracile nucleus. *Journal of Physiology* 139, 385–399.

Gordon, J.K., Matthaei, C. and van Heezik, Y. (2010) Belled collars reduce catch of domestic cats in New Zealand by half. *Wildlife Research* 37, 372–378.

Goulet, S., Dore, F.Y. and Rousseau, R. (1994) Object permanence and working memory in cats (*Felis catus*). *Journal of Experimental Psychology: Animal Behavior Processes* 20, 347–365.

Grastyan, E. and Vereczkei, L. (1974) Effects of spatial separation of the conditioned signal from the reinforcement: a demonstration of the conditioned character of the orienting response or the orientational character of conditioning. *Behavioural Biology* 10, 121–146.

Gray, J.A.B. (1966) The representation of information about rapid changes in a population of receptor units signalling mechanical events. In: de Reuck, A.V.S. and Knight, J. (eds) *Touch, Heat and Pain*. CIBA Foundation, London, pp. 299–315.

Grillner, S., Wallén, P., Saitoh, K., Kozlov, A. and Robertson, B. (2008) Neural bases of goal-directed locomotion in vertebrates—an overview. *Brain Research Reviews* 57, 2–12.

Gunn-Moore, D.A. (2011) Cognitive dysfunction in cats: clinical assessment and management. *Topics in Companion Animal Medicine* 26, 17–24.

Gunter, R. (1951) Visual size constancy in the cat. *British Journal of Psychology* 42, 288–293.

Gunther, I., Finkler, H. and Terkel, J. (2011) Demographic differences between urban feeding groups of neutered and sexually intact free-roaming cats following a trap–neuter–return procedure. *Journal of the American Veterinary Medical Association* 238, 1134–1140.

Guyot, G.W., Cross, H.A. and Bennett, T.L. (1980) Early social isolation of the domestic cat: responses to separation from social and nonsocial rearing stimuli. *Developmental Psychobiology* 13, 309–315.

Hall, S.L. (1998) Object play by adult animals. In: Bekoff, M. and Byers, J.A. (eds) *Animal Play: Evolutionary, Comparative and Ecological Perspectives*. Cambridge University Press, Cambridge, pp. 45–60.

Hall, S.L. and Bradshaw, J.W.S. (1998) The influence of hunger on object play by adult domestic cats. *Applied Animal Behaviour Science* 58, 143–150.

Hall, S.L., Bradshaw, J.W.S. and Robinson, I.H. (2002) Object play in adult domestic cats: the roles of habituation and disinhibition. *Applied Animal Behaviour Science* 79, 263–271.

Hart, B.L. (1974) Normal behavior and behavioral problems associated with sexual function, urination and defecation. *Veterinary Clinics of North America* 4, 589–606.

Hart, B.L. (1979) Breed-specific behaviour. *Feline Practice* 9, 10–13.

Hart, B.L. and Barrett, R.E. (1973) Effects of castration on fighting, roaming and urine spraying in adult male cats. *Journal of the American Veterinary Medical Association* 163, 290–292.

Hart, B.L. and Leedy, M.G. (1987) Stimulus and hormonal determinants of flehmen behaviour in cats. *Hormones and Behaviour* 21, 44–52.

Haskins, R. (1977) Effect of kitten vocalizations on maternal behavior. *Journal of Comparative and Physiological Psychology* 91, 830–838.

Haskins, R. (1979) A causal analysis of kitten vocalization: an observational and experimental study. *Animal Behaviour* 27, 726–736.

Heath, S.E. (2005) Behaviour problems and welfare. In: Rochlitz, I. (ed.) *The Welfare of Cats*. Springer, Dordrecht, The Netherlands, pp. 91–118.

Heffner, H.E. (1998) Auditory awareness. *Applied Animal Behaviour Science* 57, 259–268.

Heffner, R.S. and Heffner, H.E. (1985) Hearing range of the domestic cat. *Hearing Research* 19, 85–88.

Hein, A. and Held, R. (1967) Dissociation of the visual placing response into elicited and guided components. *Science* 158, 390–391.

Held, S. and Spinka, M. (2011) Animal play and animal welfare. *Animal Behaviour* 81, 891–899.

Hendriks, W.H., Moughan, P.J., Tarttelin, M.F. and Woolhouse, A.D. (1995) Felinine: a urinary amino acid of Felidae. *Comparative Biochemistry and Physiology B* 112, 581–588.

Herman, M.D., Denlinger, S.L., Patarca, R., Katz, L. and Hobson, J.A. (1991) Developmental phases of sleep and motor behaviour in a cat mother–infant system: a time-lapse video approach. *Canadian Journal of Psychology* 45, 101–114.

Herron, M.E. (2010) Advances in understanding and treatment of feline inappropriate elimination. *Topics in Companion Animal Medicine* 25, 195–202.

Hewson-Hughes, A.K., Hewson-Hughes, V.L., Miller, A.T., Hall, S.R., Simpson, S.J. and Raubenheimer, D. (2011) Geometric analysis of macronutrient selection in the adult domestic cat, *Felis catus*. *Journal of Experimental Biology* 214, 1039–1051.

Hobhouse, L.T. (1915) *Mind in Evolution*, 2nd Edn. MacMillan, London.

Horn, J., Mateus-Pinilla, N., Warner, R.E. and Heske, E. (2011) Home range, habitat use, and activity patterns of free-roaming domestic cats. *The Journal of Wildlife Management* 75, 1177–1185.

Houpt, K.A (2011) *Domestic Animal Behavior for Veterinarians and Animal Scientists*. Wiley-Blackwell, Ames, Iowa.

Houpt, W.J. and Wolski, T.R. (1982) *Domestic Animal Behaviour for Veterinarians and Animal Scientists*. Iowa State University Press, Ames, Iowa.

Howland, D., Bregman, B. and Glodberger, M. (1995) The development of quadrupedal locomotion in the kitten. *Experimental Neurobiology* 135, 93–107.

Huang, A.Y. and May, B.J. (1996) Sound orientation behavior in cats. II. Mid-frequency spectral cues for sound localization. *Journal of the Acoustical Society of America* 100, 1070–1080.

Huang, G.T., Rosowski, J.J. and Peake, W.T. (2000) Relating middle-ear acoustic performance to body size in the cat family: measurements and models. *Journal of Comparative Physiology A* 186, 447–465.

Hubel, D.H. and Wiesel, T.N. (1971) Aberrant visual projections in the Siamese cat. *Journal of Physiology* 218, 33–62.

Hudson, R., Raihani, G., Gonzalez, D., Bautista, A. and Distel, H. (2009) Nipple preference and contests in suckling kittens of the domestic cat are unrelated to presumed nipple quality. *Developmental Psychobiology* 51, 322–332.

Hudson, R., Guarneros, M., Rodriguez, A. and Raihani, G. (2010) Emergence of personality in kittens: a qualitative approach to detecting individual behavioural differences among littermates during the first postnatal month (abstract). *Developmental Psychobiology* 52, 703.

Hughes, A. (1972) Vergence in the cat. *Vision Research*, 12, 1961–1964.

Hughes, A. (1977) The topography of vision in mammals of contrasting life style: comparative optics and retinal organisation. In: Crescitelli, F. (ed.) *The Visual System in Vertebrates*. Springer-Verlag, New York, pp. 613–756.

Iggo, A. (1966) Cutaneous receptors with a high sensitivity to mechanical displacement. In: de Reuck, A.V.S. and Knight, J. (eds) *Touch, Heat and Pain*. CIBA Foundation, London, pp. 237–256.

Iggo, A. (1982) Cutaneous sensory mechanisms. In: Barlow, H.B. and Mollon, J.D. (eds) *The Senses*. Cambridge University Press, Cambridge, pp. 369–408.

Ishida, Y. and Shimizu, M. (1998) Influence of social rank on defecating behaviors in feral cats. *Journal of Ethology* 16, 15–21.

Ishida, Y., Yahara, T., Kasuya, E. and Yamane, A. (2001) Female control of paternity during copulation: inbreeding avoidance in feral cats. *Behaviour* 138, 235–250.

Jalowiec, J.E., Panksepp, J., Shabshelowitz, H., Zolovick, A.J., Stern, W. and Morgane, P.J. (1973) Suppression of feeding in cats following 2-deoxy-D-glucose. *Physiology and Behavior* 10, 805–807.

Jerison, H.J. (1985) Animal intelligence as encephalisation. *Philosophical Transactions of the Royal Society of London B* 308, 21–35.

Jessup, D.A. (2004) The welfare of feral cats and wildlife. *Journal of the American Veterinary Medical Association* 225, 1377–1383.

John, E.R., Chesler, P., Bartlett, F. and Victor, I. (1968) Observational learning in cats. *Science* 159, 1489–1491.

Johnson, R.F., Randall, S. and Randall, W. (1983) Freerunning and entrained circadian rhythms in activity, eating and drinking in the cat. *Journal of Interdisciplinary Cycle Research* 14, 315–327.

Johnson, R.W. (2002) The concept of sickness behaviour: a brief chronological account of four key discoveries. *Veterinary Immunology and Immunopathology* 87, 443–450.

Jones, A.K., Kulkani, B. and Derbyshire, S.W. (2003) Pain mechanisms and their disorders. *British Medical Bulletin* 65, 83–93.

Jongman, E.C. (2007) Adaptation of domestic cats to confinement. *Journal of Veterinary Behavior* 2, 193–196.

Kalkstein, T.S., Kruger, J.M. and Osborne, C.A. (1999) Feline lower urinary tract disease. Part I. Clinical manifestations. *Compendium of Continuing Education for Practicing Vets* 21, 15–26.

Kane, E. (1989) Feeding behaviour of the cat. In: Burger, I.H. and Rivers, J.P.W. (eds) *Nutrition of the Dog and Cat*. Cambridge University Press, Cambridge, pp. 147–158.

Karsh, E.B. (1983) The effects of early and late handling on the attachment of cats to people. In: Anderson, R.K., Hart, B.L. and Hart, L.A. (eds) *The Pet Connection, Conference Proceedings*. Globe Press, St Paul, Minnesota.

Karsh, E.B. and Turner, D.C. (1988) The human–cat relationship. In: Turner, D.C. and Bateson, P. (eds) *The Domestic Cat: The Biology of its Behaviour*. Cambridge University Press, Cambridge, pp. 159–177.

Kays, R. and DeWan, A. (2004) Ecological impact of inside/outside house cats around a suburban nature preserve. *Animal Conservation* 7, 273–283.

Kerby, G. and Macdonald, D.W. (1988) Cat society and the consequences of colony size. In: Turner, D.C. and Bateson, P. (eds) *The Domestic Cat: The Biology of its Behaviour*. Cambridge University Press, Cambridge, pp. 67–81.

Kessler, M.R. and Turner, D.C. (1997) Stress and adaptation of cats (*Felis silvestris catus*) housed singly, in pairs and in groups in boarding catteries. *Animal Welfare* 6, 243–254.

Kessler, M.R. and Turner, D.C. (1999) Socialization and stress in cats (*Felis silvestris catus*) housed singly and in groups in animal shelters. *Animal Welfare* 8, 15–26.

Kiley-Worthington, M. (1976) The tail movements of ungulates, canids and felids with particular reference to their causation and function as displays. *Behaviour* 56, 69–115.

Kiley-Worthington, M. (1984) Animal language? Vocal communication of some ungulates, canids and felids. *Acta Zoologica Fennica* 171, 83–88.

Kimyai-Asadi, A. and Usman, A. (2001) The role of psychological stress in skin disease. *Journal of Cutaneous Medicine and Surgery* 5, 140–145.

Kitchener, A. (1991) *The Natural History of the Wild Cats*. Christopher Helm, London.

Knowles, R.J., Curtis, T.M. and Crowell-Davis, S.L. (2004) Correlation of dominance as determined by agonistic interactions with feeding order in cats. *American Journal of Veterinary Research* 65, 1548–1556.

Kolb, B. and Nonneman, A.J. (1975) The development of social responsiveness in kittens. *Animal Behaviour* 23, 368–374.

Kry, K. and Casey, R. (2007) The effect of hiding enrichment on stress levels and behaviour of domestic cats (*Felis sylvestris catus*) in a shelter setting and the implications for adoption potential. *Animal Welfare* 16, 375–383.

Kulkarni, B., Bentley, D.E., Elliot, R., Youell, P., Watson, A. and Derbyshire, S.W. (2005) Attention to pain localisation and unpleasantness discriminates the functions of the medial and lateral pain systems. *European Journal of Neuroscience* 21, 3133–3142.

Kuo, Z.Y. (1960) Studies on the basic factors in animal fighting. VII. Interspecies coexistence in mammals. *Journal of Genetic Psychology* 97, 211– 225.

Landsberg, G., Denenberg, S. and Araujo, J. (2010) Cognitive dysfunction in cats. A syndrome we used to dismiss as 'old age'. *Journal of Feline Medicine and Surgery* 12, 837–848.

Lascelles, D. and Robertson, S. (2010) DJD-associated pain in cats: what can we do to promote patient comfort? *Journal of Feline Medicine and Surgery* 12, 200–212.

Levine, M.S., Lloyd, R.L, Fisher, R.S., Hull, C.D. and Buchwald, N.A. (1987) Sensory, motor and cognitive alterations in aged cats. *Neurobiology of Aging* 8, 253–263.

Leyhausen, P. (1979) *Cat Behavior: The Predatory and Social Behavior of Domestic and Wild Cats*. Garland STPM Press, New York.

Li, X., Li, W., Wang, H., Cao, J., Maehashi, K., Huang, L. *et al.* (2005) Pseudogenization of a sweet-receptor gene accounts for cats' indifference toward sugar. *PLoS Genetics* 1, e3.

Liberg, O. and Sandell, M. (1988) Spatial organisation and reproductive tactics in the domestic cat and other felids. In: Turner, D.C. and Bateson, P. (eds) *The Domestic Cat: The Biology of its Behaviour*. Cambridge University Press, Cambridge, pp. 83–98.

Liberg, O., Sandell, M., Pontier, D. and Natoli, E. (2000) Density, spatial organization and reproductive tactics in the domestic cat and other felids. In: Turner, D.C. and Bateson, P. (eds) *The Domestic Cat: The Biology of its Behaviour*, 2nd Edn. Cambridge University Press, Cambridge, pp. 120–147.

Lilith, M., Calver, M. and Garkaklis, M. (2008) Roaming habits of pet cats on the suburban fringe in Perth, Western Australia – what size buffer zone is needed to protect wildlife in reserves? *Conference Proceedings: Too Close for Comfort: Contentious Issues in Human–Wildlife Encounters*, pp. 65–72.

Lilith, M., Calver, M. and Garkaklis, M. (2010) Do cat restrictions lead to increased species diversity or abundance of small and medium-sized mammals in remnant urban bushland? *Pacific Conservation Biology* 16, 162–172.

Lipinski, M.J., Froenicke, L., Baysac, K.C., Billings, N.C., Leutenegger, C.M., Levy, A.M. *et al.* (2008) The ascent of cat breeds: genetic evaluations of breeds and worldwide random-bred populations. *Genomics* 91, 12–21.

Loffler, G. (2008) Perception of contours and shapes: low and intermediate stage mechanisms. *Vision Research* 48, 2106–2127.

Longcore, T., Rich, C. and Sullivan, L.M. (2009) Critical assessment of claims regarding management of feral cats by trap–neuter–return. *Conservation Biology* 23, 887–894.

Loop, M.S. and Frey, T.J. (1981) Critical flicker fusion in Siamese cats. *Experimental Brain Research* 43, 65–68.

Loop, M.S., Millican, C.L. and Thomas, S.R. (1987) Photopic spectral sensitivity of the cat. *Journal of Physiology* 382, 537–553.

Lowe, S. and Bradshaw, J.W.S. (2001) Ontogeny of individuality in the domestic cat in the home environment. *Animal Behaviour* 61, 231–237.

Lowe, S. and Bradshaw, J.W.S. (2002) Responses of pet cats to being held by an unfamiliar person. *Anthrozoos* 15 69–69.

Macdonald, D.W. (1983) The ecology of carnivore social behaviour. *Nature* 301, 379–389.

Macdonald, D.W. and Loveridge, A.J. (eds) (2010) *Biology and Conservation of Wild Felids.* Oxford University Press, Oxford.

Macdonald, D.W., Apps, P.J., Carr, G.M. and Kerby, G. (1987) Social dynamics, nursing coalitions and infanticide among farm cats, *Felis catus*. *Advances in Ethology (supplement to Ethology)* 28, 1–64.

Macdonald, D.W., Yamaguchi, N. and Kerby, G. (2000) Group-living in the domestic cat: its sociobiology and epidemiology. In: Turner, D.C. and Bateson, P. (eds) *The Domestic Cat: The Biology of its Behaviour*, 2nd Edn. Cambridge University Press, Cambridge, pp. 95–118.

MacDonald, M.L., Rogers, Q.R. and Morris, J.G. (1985) Aversion of the cat to dietary medium-chain triglycerides and caprylic acid. *Physiology and Behavior* 35, 371–375.

MacDonnell, M.F. and Flynn, J.P. (1966) Control of sensory fields by stimulation of hypothalamus. *Science* 152, 1406–1408.

Malmström, T. and Kröger, R.H.H. (2006) Pupil shapes and lens optics in the eyes of terrestrial vertebrates. *Journal of Experimental Biology* 209, 18–25.

Marchei, P., Divero, S., Falocci, Fatjo, J., Ruiz-de-la-Torre, J.L. and Manteca, X. (2009) Breed differences in behavioural development in kittens. *Physiology and Behavior* 96, 522–531.

Martin, P. (1984) The (four) whys and wherefores of play in cats: a review of functional, evolutionary, developmental and causal issues. In: Smith, P.K. (ed.) *Play in Animals and Humans*. Blackwell, Oxford, UK, pp. 71–94.

Martin, P. and Bateson, P. (1985) The ontogeny of locomotor play behaviour in the domestic cat. *Animal Behaviour* 33, 502–510.

Martin, P. and Bateson, P. (1988) Behavioural development in the cat. In: Turner, D.C. and Bateson, P. (eds) *The Domestic Cat: The Biology of its Behaviour*. Cambridge University Press, Cambridge, pp. 9–22.

Martin, R.L. and Webster, W.R. (1987) The auditory spatial acuity of the domestic cat in the interaural horizontal and median vertical planes. *Hearing Research* 30, 239–252.

Martin, R.L. and Webster, W.R. (1989) Interaural sound pressure level differences associated with sound-source locations in the frontal hemifield of the domestic cat. *Hearing Research* 38, 289–302.

Matzner, O. and Devor, M. (1992) Na+ conductance and the threshold for repetitive neuronal firing. *Brain Research* 597, 92–98.

McCann, S.M., Minura, M., Karanth, S., Yu, W.H., Mastronardi, C.A. and Rettori, V. (2000) The mechanism of action of cytokines to control the release of hypothalamic and pituitary hormones in infection. *Annals of the New York Academy of Sciences* 917, 4–18.

McCobb, E.C., Patronek, G.J., Marder, A., Dinnage, J.D. and Stone, M.S. (2005) Assessment of stress levels among cats in four animal shelters. *Journal of the American Veterinary Medical Association* 226, 548–555.

McComb, K., Taylor, A.M., Wilson, C. and Charlton, B.D. (2009) The cry embedded within the purr. *Current Biology* 19, R507–R508.

McCune, S. (1992) Temperament and the welfare of caged acts. PhD thesis, University of Cambridge, Cambridge, UK.

McCune, S. (1995) The impact of paternity and early socialization on the development of cats' behaviour to people and novel objects. *Applied Animal Behaviour Science* 45, 109–124.

McCune, S., McPherson, J. and Bradshaw, J. (1995) Avoiding problems: the importance of socialization. In: *The Waltham Book of Human–Animal Interaction: Benefits and Responsibilities of Pet Ownership*. Pergamon/Elsevier Science, Ltd, Oxford, UK.

McCune, S., Stevenson, J., Fretwell, L., Thompson, A. and Mills, D.S. (2008) Ageing does not significantly affect performance in a spatial learning task in the domestic cat (*Felis sylvestris catus*). *Applied Animal Behaviour Science* 112, 345–356.

McEllistrem, J.E. (2004) Affective and predatory violence: a bimodal classification system of human aggression and violence. *Aggression and Violent Behavior* 10, 1–30.

McFarland, D. (1985) *Animal Behaviour: Psychobiology, Ethology and Evolution*. Longman, Harlow, UK.

McGreevy, P.D. and Boakes, R.A. (2007) *Carrots and Sticks: Principles of Animal Training*. Cambridge University Press, Cambridge, p. 318.

McVea, D.A. and Pearson, K.G. (2007) Stepping of the forelegs over obstacles establishes longlasting memories in cats. *Current Biology* 17, R621–R623.

Medina, F.M., Bonnaud, E., Vidal, E., Tershy B., Zavaleta, E.S., Donlan, C.J. *et al.* (2011) A global review of the impacts of invasive cats on island endangered vertebrates. *Global Change Biology* 17, 3503–3510.

Meek, P. (2003) Home range of house cats (*Felis catus*) living within a national park. *Australian Mammalogy* 25, 51–60.

Meier, M. and Turner, D.C. (1985) Reactions of home cats during encounters with a strange person: evidence for two personality types. *Journal of the Delta Society (later Antrozoos)* 2, 45–53.

Mellen, J.D. (1988) The effects of hand-raising on sexual behavior of captive small felids using domestic cats as a models. *Annual Proceedings of the American Association of Zoological Parks and Aquariums*, pp. 253–259.

Mellen, J.D. (1993) A comparative analysis of scent-marking, social and reproductive behaviour in 20 species of small cats (*Felis*). *American Zoologist* 33, 151–166.

Mendl, M. (1988) The effects of litter-size variation on the development of play behaviour in the domestic cat: litters of one and two. *Animal Behaviour* 36, 20–34.

Mendl, M. and Harcourt, R. (1988) Individuality in the domestic cat. In: Turner, D.C. and Bateson, P. (eds) *The Domestic Cat: The Biology of its Behaviour*. Cambridge University Press, Cambridge, pp. 41–54.

Mendl, M., Brooks, J., Basse, C., Burman, O., Paul, E., Blackwell, E. *et al.* (2010) Dogs showing separation-related behaviour exhibit a 'pessimistic' cognitive bias. *Current Biology* 20, 839–840.

Mendoza, D.L. and Ramirez, J.M. (1987) Play in kittens (*Felis domesticus*) and its association with cohesion and aggression. *Bulletin of the Psychonomic Society* 25, 27–30.

Menotti-Raymond, M.A. and O'Brien, S. (1995) Evolutionary conservation of ten microsatellite loci in four species of Felidae. *Journal of Heredity* 86, 319–322.

Mertens, C. (1991) Human–cat interactions in the human setting. *Anthrozoos* 4, 214–231.

Mertens, C. and Schär, R. (1988) Practical aspects of research on cats. In: Turner, D.C. and Bateson, P. (eds) *The Domestic Cat: the Biology of its Behaviour*. Cambridge University Press, Cambridge, pp. 179–190.

Mertens, C. and Turner, D.C. (1988) Experimental analysis of human–cat interactions during first encounters. *Anthrozoos* 2, 83–97.

Metsers, E., Seddon, P. and van Heezik, Y. (2010) Cat-exclusion zones in rural and urban-fringe landscapes: how large would they have to be? *Wildlife Research* 37, 47–56.

Michael, R.P. (1961) Observations upon the sexual behaviour of the domestic cat (*Felis catus* L.) under laboratory conditions. *Behaviour* 18, 1–24.

Miklósi, A. (2007) Human–animal interactions and social cognition in dogs. In: Jensen, P. (ed.) *The Behavioural Biology of Dogs*. CAB International, Wallingford, UK, pp. 207–222.

Miklósi, A., Pongracz, P., Lakatos, G., Topal, J. and Csanyi, V. (2005) A comparative study of the use of visual communicative signals in interactions between dogs and humans and cats and humans. *Journal of Comparative Psychology* 119, 179–186.

Miles, R.C. (1958) Learning in kittens with manipulatory, exploratory and food incentives. *Journal of Comparative and Physiological Psychology* 51, 39–42.

Miyazaki, M., Yamashita, T., Taira, H. and Suzuki, A. (2008) The biological function of cauxin, a major urinary protein of the domestic cat (*Felis catus*). In: Hurst, J.L., Beynon, R.J., Roberts, S.C. and Wyatt, T. (eds) *Chemical Signals in Vertebrates II*. Springer, New York, pp. 51–60.

Moelk, M. (1944) Vocalizing in the house-cat; a phonetic and functional study. *American Journal of Psychology* 57, 184–205.

Morgan, S.A., Hansen, C.M., Ross, J.G., Hickling, G.J. and Ogilvie, S.C. (2009) Urban cat (*Felis catus*) movement and predation activity associated with a wetland reserve in New Zealand. *Wildlife Research* 36, 574–580.

Morris, J.G. (2002) Idiosyncratic nutrient requirements of cats appear to be diet-induced evolutionary adaptations. *Nutrition Research Reviews* 15, 153–168.

Mugford, R.A. (1977) External influences on the feeding of carnivores. In: Kare, M.R. and Maller, O. (eds) *The Chemical Senses and Nutrition*. Academic Press, New York, pp. 25–50.

Mumma, R. and Warren, J.M. (1968) Two-cue discrimination learning by cats. *Journal of Comparative and Physiological Psychology* 66, 116–121.

Murray, C.D., Flynn, J., Ratcliffe, L., Jacyna, M.R., Kamm, M.A. and Emmanuel, A.V. (2004) Acute physical and psychological stress and its influence on autonomic outflow to the gut in irritable bowel syndrome. *Gut* 53(Suppl. 3), A29.

Murray, J.K., Browne, W.J., Roberts, M.A., Whitmarsh, A. and Gruffydd-Jones, T.J. (2010) Number and ownership profiles of cats and dogs in the UK. *Veterinary Record* 166, 163–168.

Nater, U.M., Gaab, J., Rief, W. and Ehlert, U. (2006) Recent trends in behavioral medicine. *Current Opinion in Psychiatry* 19, 180–183.

Natoli, E. (1985) Behavioural responses of urban feral cats to different types of urine marks. *Behaviour* 94, 234–243.

Natoli, E. (1990) Mating strategies in cats: a comparison of the role and importance of infanticide in domestic cats, *Felis catus* L., and lions, *Panthera leo* L. *Animal Behaviour* 40, 183–186.

Natoli, E. and De Vito, E. (1991) Agonistic behaviour, dominance rank and copulatory success in a large multi-male feral cat colony (*Felis catus*) in central Rome. *Animal Behaviour* 42, 227–241.

Natoli, E., De Vito, E. and Pontier, D. (2000) Mate choice in the domestic cat (*F. catus*) *Aggressive Behavior* 26, 455–465.

Natoli, E., Baggio, B. and Pontier, D. (2001) Male and female agonistic and affiliative relationships in a social group of farm cats (*Felis catus*). *Behavioural Processes* 53, 137–143.

Natoli, E., Schmid, M., Say, L. and Pontier, D. (2007) Male reproductive success in a social group of urban feral cats (*Felis catus L.*). *Ethology* 113, 283–289.

Nelson, S.H., Evans, A.D. and Bradbury, R.B. (2005) The efficacy of collar-mounted devices in reducing the rate of predation of wildlife by domestic cats. *Applied Animal Behaviour Science* 94, 273–285.

Nicastro, N. (2004) Perceptual and acoustic evidence for species-level differences in meow vocalizations by domestic cats (*Felis catus*) and African wild cats (*Felis sylvestris lybica*). *Journal of Comparative Psychology* 118, 287–296.

Nicastro, N. and Owren, M.J. (2003) Classification of domestic cat (*Felis catus*) vocalizations by naive and experienced human listeners. *Journal of Comparative Psychology* 117, 44–52.

Nieder, A. (2002) Seeing more than meets the eye: processing of illusory contours in animals. *Journal of Comparative Physiology A* 188, 249–260.

Nogales, M., Martin, A., Tershy, B.R., Donlan, J., Veitch, D., Puerta, N. *et al.* (2004) A review of feral cat eradication on islands. *Conservation Biology* 18, 310–319.

Norton, T.T. (1974) Receptive-field properties of superior colliculus cells and development of visual behaviour in kittens. *Journal of Neurophysiology* 37, 674–690.

O'Brien, S.J. and Johnson, W.E. (2007) The evolution of cats. Genomic paw prints in the DNA of the world's wild cats have clarified the cat family tree and uncovered several remarkable migrations in their past. *Scientific American* 297, 68–75.

O'Brien, S.J., Johnson, W., Driscoll, C., Pontius, J., Pecon-Slattery, J. and Menotti-Raymond, M. (2008) State of cat genomics. *Trends in Genetics* 24, 268–279.

O'Haire, M. (2010) Companion animals and human health: benefits, challenges, and the road ahead. *Journal of Veterinary Behavior* 5, 226–234.

Olmstead, C.E. and Villablanca, J.R. (1980) Development of behavioural audition in the kitten. *Physiology and Behaviour* 24, 705–712.

Oswald, I. (1962) *Sleeping and Waking*. Elsevier, Amsterdam.

Ottway, D.S. and Hawkins, D.M. (2003) Cat housing in rescue shelters: a welfare comparison between communal and discrete-unit housing. *Animal Welfare* 12, 173–177.

Pageat, P. and Gaultier, E. (2003) Current research in canine and feline pheromones. *Veterinary Clinics of North America: Small Animal Practice* 33, 187–211.

Palen, G.F. and Goddard, G.V. (1966) Catnip and oestrous behaviour in the cat. *Animal Behaviour* 14, 372–377.

Passanisi, W.C. and Macdonald, D.W. (1990) Group discrimination on the basis of urine in a farm cat colony. In: Macdonald, D.W., Muller-Schwarze, D. and Natynczuk, S.E. (eds) *Chemical Signals in Vertebrates 5*. Oxford University Press, Oxford, pp. 336–345.

Pasternak, T. and Merigan, W.H. (1980) Movement detection by cats: invariance with direction and target configuration. *Journal of Comparative and Physiological Psychology* 94, 943–952.

Perfiliev, S., Pettersson, L.G. and Lundberg, A. (1998) Control of claw movements in cats. *Neuroscience Research* 31, 337–342.

PETA (2011) Inside the Fur Industry: Factory Farms. People for the Ethical Treatment of Animals, Norfolk, Virginia. Available at: http://www.peta.org/issues/Animals-Used-for-Clothing/inside-the-fur-industry-factory-farms.aspx. (accessed 28 August 2012).

Plantinga, E.A., Bosch, G. and Hendriks, W.H. (2011) Estimation of the dietary nutrient profile of free-roaming feral cats: possible implications for nutrition of domestic cats. *British Journal of Nutrition* 106, S35–S48.

Podberscek, A.L. (2009) Good to pet and eat: the keeping and consuming of dogs and cats in South Korea. *Journal of Social Issues* 65, 615–632.

Pontier, D. and Natoli, E. (1996) Reproductive success of male domestic cats (*Felis catus*): a case history. *Behavioural Processes* 37, 85–88.

Pontier, D. and Natoli, E. (1999) Infanticide in rural male cats (*Felis catus*) as a reproductive mating tactic. *Aggressive Behaviour* 25, 445–449.

Poucet, B. (1985) Spatial behaviour of cats in cue-controlled environments. *Quarterly Journal of Experimental Psychology* 37B, 155–179.

Pozza, M.E., Stella, J.L., Chappuis-Gagnon, A.-C., Wagner, S.O. and Buffington, C.A.T. (2008) Pinch-induced behavioral inhibition ('clipnosis') in domestic cats. *Journal of Feline Medicine and Surgery* 10, 82–87.

Pryor, P.A., Hart, B.L., Bain, M.J. and Cliff, K.D. (2001) Causes of urine marking in cats and effects of environmental management on the frequency of marking. *Journal of the American Veterinary Medical Association* 219, 1709–1713.

Radinsky, L. (1975) Evolution of the felid brain. *Brain Behaviour & Evolution* 11, 214–254.

Raihani, G., Gonzalez, D., Arteaga, L. and Hudson, R. (2009) Olfactory guidance of nipple attachment and suckling in kittens of the domestic cat: inborn and learned responses. *Developmental Psychobiology* 51, 662–671.

Rainville, P. (2002) Brain mechanisms of pain affect and pain modulation. *Current Opinion in Neurobiology* 12, 195–204.

Randall, W. and Parsons, V. (1987) Three views of the annual phase map of the domestic cat, *Felis catus* L. *Journal of Interdisciplinary Cycle Research* 18, 17–28.

Reed, D.R., Tanaka, T. and McDaniel, A.H. (2006) Diverse tastes: genetics of sweet and bitter perception. *Physiology & Behavior* 88, 215–226.

Reis, P.M., Jung, S., Aristoff, J.M. and Stocker, R. (2010) How cats lap: water uptake by *Felis catus*. *Science* 330, 1231–1234.

Reisner, I. (1991) The pathophysiologic basis of behavior problems. *Veterinary Clinics of North America: Small Animal Practice* 21, 207– 224.

Reisner, I., Houpt, K., Erb, H. and Quimby, F. (1994) Friendliness to humans and defensive aggression in cats: the influence of handling and paternity. *Physiology and Behaviour* 55, 1119–1124.

Rizzo, M.A., Kocsis, J.D. and Waxman, S.G. (1996) Mechanisms of paresthesiae, dysesthesiae and hyperesthesiae: role of Na+ channel heterogeneity. *European Neurology* 36, 3–12.

Robertson, S.A. (2008) A review of feral cat control. *Journal of Feline Medicine and Surgery* 10, 366–375.

Robertson, S.A. and Lascelles, B.D.X. (2010) Long-term pain in cats: how much do we know about this important welfare issue? *Journal of Feline Medicine and Surgery* 12, 188–199.

Rochlitz, I. (ed.) (2005a) *The Welfare of Cats*. Springer, Dordrecht, The Netherlands.

Rochlitz, I. (2005b) Housing and welfare. In: Rochlitz, I. (ed.) *The Welfare of Cats*. Springer, Dordrecht, The Netherlands, pp. 177–203.

Romand, R. and Ehret, G. (1984) Development of sound production in normal, isolated, and deafened kittens during the first postnatal months. *Developmental Psychobiology* 17, 629–649.

Rosenblatt, J.S. (1972) Learning in newborn kittens. *Scientific American* 227, 18–25.

Rossignol, R., Dubuc, R. and Gossard J.-P. (2006) Dynamic sensorimotor interactions in locomotion. *Physiological Reviews* 86, 89–154.

Rowan, A.N. and Williams, J. (1987) The success of companion animal management programs: a review. *Anthrozoos* 1, 110–122.

Rozin, P. (1976) The selection of foods by rats, humans, and other animals. *Advances in the Study of Behaviour* 6, 21–76.

Rupprecht, R. and Holsboer, F. (1999) Neuroactive steroids: mechanisms of action and neuropsychopharmacological perspectives. *Trends in Neurosciences* 22, 410–416.

Rusbridge, C., Heath, S., Gunn-Moore, D.A., Knowler, S.P., Johnson, N. and McFayden, A.K. (2010) Feline orofacial pain syndrome (FOPS): a retrospective study of 113 cases. *Journal of Feline Medicine and Surgery* 12, 498–508.

Russell, K., Sabin, R., Holt, S., Bradley, R. and Harper, E.J. (2000) Influence of feeding regimen on body condition in the cat. *Journal of Small Animal Practice* 41, 12–17.

Ruxton, G.D., Thomas, S. and Wright, J.W. (2002) Bells reduce predation of wildlife by domestic cats (*Felis catus*). *Journal of Zoology* 256, 81–83.

Salazar, I., Quinteiro, P.S., Cifuentes, J.M. and Caballero, T.G. (1996) The vomeronasal organ of the cat. *Journal of Anatomy* 188, 445–454.

Salisbury, S.K. (1991) Hyperthyroidism in cats. *Compendium of Continuing Veterinary Education (Europe)* 13, 606–614.

Sandell, M. (1989) The mating tactics and spacing patterns of solitary carnivores. In: Gittleman, J.L. (ed.) *Carnivore Behaviour, Ecology, and Evolution 1*. Cornell University Press, New York, pp. 164–182.

Say, L. and Pontier, D. (2004) Spacing pattern in a social group of stray cats: effects on male reproductive success. *Animal Behaviour* 68, 175–180.

Say, L., Pontier, D. and Natoli, E. (1999) High variation in multiple paternity of domestic cats in relation to environmental conditions. *Proceedings of the Royal Society of London Series B* 266, 2071–2074.

Say, L., Pontier, D. and Natoli, E. (2001) Influence of oestrus synchronization on male reproductive success in the domestic cat (*F. catus*). *Proceedings of the Royal Society of London Series B* 268, 1049–1053.

Say, L., Devillard, S. and Natoli, E. (2002) The mating system of feral cats in a sub-antarctic environment. *Polar Biology* 25, 838–842.

Scherk, M. (2010) Snots and snuffles: rational approach to chronic feline upper respiratory syndromes. *Journal of Feline Medicine and Surgery* 12, 548–557.

Schmidt, P., Lopez, R. and Collier, B. (2007) Survival, fecundity and movements of free-roaming cats. *Journal of Wildlife Management* 71, 915–919.

Schneirla, T.C. and Rosenblatt, J.S. (1961) Behavioral organisation and genesis of the social bond in insects and mammals. *Journal of Orthopsychiatry* 31, 223–253.

Seawright, A., Casey, R.A., Kiddie, J., Murray, J., Gruffydd-Jones, T., Harvey, A. *et al.* (2008) A case of recurrent feline idiopathic cystitis: the control of clinical signs with behaviour therapy. *Journal of Veterinary Behavior: Clinical Applications and Research* 3, 32–38.

Serpell, J. (2000) Domestication and history of the cat. In: Turner, D.C. and Bateson, P. (eds) *The Domestic Cat: The Biology of its Behaviour*, 2nd Edn. Cambridge University Press, Cambridge, pp. 179–192.

Serpell, J.A. and Paul, E.S. (2011) Pets in the family: an evolutionary perspective. In: Salmon, C.A and Shackelford, T.K. (eds) *The Oxford Handbook of Evolutionary Family Psychology*. Oxford University Press, New York, pp. 297–309.

Shell, L.G. (1994) Feline hyperaesthesia syndrome. *Feline Practice* 22, 10.

Sheppard, G. and Mills, D.S. (2003) Construct models in veterinary behavioural medicine: lessons from the human experience. *Veterinary Research Communications* 27, 175–191.

Shettleworth, S.J. (1984) Natural history and evolution of learning in nonhuman mammals. In: Marler, P. and Terrace, H.S. (eds) *The Biology of Learning*. Springer-Verlag, Berlin, pp. 419–433.

Shihab, N., Bowen, J. and Volk, H.A. (2011) Behavioural changes in dogs associated with development of idiopathic epilepsy. *Epilepsy & Behavior* 21, 160–167.

Shimizu, M. (2001) Vocalizations of feral cats: sexual differences in the breeding season. *Mammal Study* 26, 85–92.

Siegel, A. and Pott, C.B. (1988) Neural substrates of aggression and flight in the cat. *Progress in Neurobiology* 31, 261–283.

Siegel, A. and Shaikh, M.B. (1997) The neural bases of aggression and rage in the cat. *Aggression and Violent Behavior* 2, 241–271.

Siegford, J.M., Wlashaw, S.O., Brunner, P. and Zanella, A.J. (2003) Validation of a temperament test for domestic cats. *Anthrozoös* 16, 332–351.

Silva-Rodriguez, E.A. and Sieving K.E. (2011) Influence of care of domestic carnivores on their predation on vertebrates. *Conservation Biology* 25, 808–815.

Sims, V. and Chin, M. (2002) Responsiveness and perceived intelligence as predictors of speech addressed to cats. *Anthrozoös* 15, 166–177.

Smith, D.F.E., Durman, K.J., Roy, D.B. and Bradshaw, J.W.S. (1994) Behavioural aspects of the welfare of rescued cats. *Journal of the Feline Advisory Bureau* 31, 25–28.

Smithers, R.H.N. (1968) Cat of the Pharaohs. *Animal Kingdom* 61, 16–23.

Soennichsen, S. and Chamove, A.S. (2002) Responses of cats to petting by humans. *Anthrozoös* 15, 258–265.

Sokolov, E.N. (1963) Higher nervous functions: the orienting reflex. *Annual Review of Physiology* 25, 545–580.

Sokolov, V.E., Naidenko, S.V. and Serbenyuk, M.A. (1996) Recognition by the European lynx of the species and sex and age of conspecific, familiar, and unfamiliar individuals according to urinary odors. *Izvestiya Akademii Nauk Seriya Bilogicheskaya* 5, 571–577.

Stein, B.E., Magalhaes-Castro, B. and Kruger, L. (1976) Relationship between visual and tactile representations in cat superior colliculus. *Journal of Neurophysiology* 39, 401–419.

Sunquist, M. and Sunquist, F. (2002) *Wild Cats of the World*. University of Chicago Press, Chicago, Illinois.

Takeuchi, Y. and Mori, Y. (2009) Behavioral profiles of feline breeds in Japan. *Journal of Veterinary Medical Science* 71, 1053–1057.

Tami, G., Martorell, A., Torre, C., Compagnucci, M. and Manteca, X. (2011a) A model to quantify the anticipatory response in cats. *Animal Welfare* 20, 191–200.

Tami, G., Torre, C., Compagnucci, M. and Manteca, X. (2011b) Interpretation of ambiguous spatial stimuli in cats. *Animal Welfare* 20, 185–189.

Tamura, T., Nakatani, K. and Yau, K.W. (1989) Light adaptation in cat retinal rods. *Science* 245, 755–758.

Tanaka, S., Tani, T., Ribot, J., O'Hashi, K. and Imamura, K. (2009) A postnatal critical period for orientation plasticity in the cat visual cortex. *PLoS One* 4, e5380.

Thorne, C.J. (1982) Feeding behaviour in the cat – recent advances. *Journal of Small Animal Practice* 23, 555–562.

Todd, N.B. (1977) Cats and commerce. *Scientific American* 237, 100–107.

Tollin, D.J., McClaine, E.M. and Yin, T.C.T. (2010) Short-latency, goal-directed movements of the pinnae to sounds that produce auditory spatial illusions. *Journal of Neurophysiology* 103, 446–457.

Tritsch, M.E. (1993) Color choice behavior in cats and the effect of changes in the color of the illuminant. *Naturwissenschaften* 80, 287–288.

Troy, J.B. and Shou, T. (2002) The receptive fields of cat retinal ganglion cells in physiological and pathological states: where we are after half a century of research. *Progress in Retinal and Eye Research* 21, 263–302.

Tscanz, B., Hegglin, D., Gloor, S. and Bontadina, F. (2011) Hunters and non-hunters: skewed predation rate by domestic cats in a rural village. *European Journal of Wildlife Research* 57, 597–602.

Turner, D.C. (1988) Cat behaviour and the human/cat relationship. *Animalis Familiaris* 3, 16–21.

Turner, D.C. (1991) The ethology of the human–cat relationship. *Swiss Archive for Veterinary Medicine* 133, 63–70.

Turner, D. (2000) The human–cat relationship. In: Turner, D. and Bateson, P. (eds) *The Domestic Cat: The Biology of its Behaviour*, 2nd Edn. Cambridge University Press, Cambridge, pp. 193–206.

Turner, D.C. and Bateson, P. (2000) Why the cat? In: Turner, D.C. and Bateson, P. (eds) *The Domestic Cat: The Biology of its Behaviour*, 2nd Edn. Cambridge University Press, Cambridge, pp. 3–6.

Turner, D.C. and Meister, O. (1988) Hunting behaviour of the domestic cat. In: Turner, D.C. and Bateson, P. (eds) *The Domestic Cat: The Biology of its Behaviour*. Cambridge University Press, Cambridge, pp. 111–121.

Turner, S. (2004) Blindness in pets: seeing solutions. *Veterinary Times* 34, 8–9.

UK Cat Behaviour Working Group (1995) *An Ethogram for Behavioural Studies of the Domestic Cat (Felis silvestris catus L)*. UFAW, Whearthampstead, UK.

Uz, T., Akhisaroglu, M., Ahmed, R. and Manev, H. (2003) The pineal gland is critical for circadian Period1 expression in the striatum and for circadian cocaine sensitization in mice. *Neuropsychopharmacology* 28, 2117–2123.

Van den Bos, R. (1998) The function of allogrooming in domestic cats (*Felis silvestris catus*); a study in a group of cats living in confinement. *Journal of Ethology* 16, 1–13.

Van den Bos, R., Meijer, M.K. and Spruijt, B.M. (2000) Taste reactivity patterns in domestic cats (*Felis silvestris catus*). *Applied Animal Behaviour Science* 69, 149–168.

Vazquez-Dominguez, E., Ceballos, G. and Cruzado, J. (2004) Extirpation of an insular subspecies by a single introduced cat: the case of the endemic deer mouse *Peromyscus guardia* on Estanque Island, Mexico. *Oryx* 38, 347–350.

Verberne, G. and De Boer, J.N. (1976) Chemocommunication among domestic cats mediated by the olfactory and vomeronasal senses. *Zeitschrift fur Tierpsychologie* 42, 86–109.

Vigne, J.-D., Guilaine, J., Debue, K., Haye, L., and Gérard, P. (2004) Early taming of the cat in Cyprus. *Science* 304, 259.

Villablanca, J.R. and Olmstead, C.E. (1979) Neurological development of kittens. *Developmental Psychobiology* 12, 101–127.

Warren, J.M. (1960) Oddity learning set in a cat. *Journal of Comparative and Physiological Psychology* 53, 433–434.

Warren, J.M. (1972) Transfer of responses to open and closed shapes in discrimination learning by cats. *Perception & Psychophysics* 12, 449–452.

Warren, J.M. (1976) Irrelevant cues and shape discrimination learning by cats. *Animal Learning and Behaviour* 4, 22–24.

Warren, J.M. and Beck, C.H. (1966) Visual probability learning by cats. *Journal of Comparative and Physiological Psychology* 61, 316–318.

Watt, D.G.D. (1976) Responses of cats to sudden falls: an otolith originating reflex assisting landing. *Journal of Neurophysiology* 39, 257–265.

Waxman, S.G., Dib-Hajj, S., Cummins, T.R. and Black, J.A. (2000) Sodium channels and their genes: dynamic expression in the normal nervous system, dysregulation in disease states. *Brain Research* 886, 5–14.

Wedl, M., Bauer, B., Gracey, D., Grabmayer, C., Spielauer, E., Day, J. *et al.* (2011) Factors affecting the temporal patterns of dyadic behaviours and interactions between domestic cats and their owners. *Behavioural Processes* 86, 58–67.

Weinstein, L. and Alexander, R. (2010) College students and their cats. *College Student Journal Publisher: Project Innovation (Alabama)* 44.

Wells, D.L. and Hepper, P.G. (1992) The behaviour of dogs in a rescue shelter. *Animal Welfare* 1, 171–186.

Wells, D.L. and Millsopp, S. (2009) Lateralised behaviour in the domestic cat, *Felis silvestris catus*. *Animal Behaviour* 78, 537–541.

West, M.J. (1974) Social play in the domestic cat. *American Zoologist* 14, 427–436.

West, M.J. (1979) Play in domestic kittens. In: Cairns, R.B. (ed.) *The Analysis of Social Interactions*. Hillsdale, New Jersey.

Wetzel, M.C., Anderson, R.C., Brady, T.H. and Norgren, K.S. (1977) Kinematics of treadmill galloping by cats. III. Coordination during gait conversions and implications for neural control. *Behavioral Biology* 21, 107–127.

White, T.D. and Boudreau, J.C. (1975) Taste preferences of the cat for neurophysiology active compounds. *Physiological Psychology* 3, 405–410.

Whitt, E., Douglas, M., Osthaus, B. and Hocking, I. (2009) Domestic cats (*Felis catus*) do not show causal understanding in a string-pulling task. *Animal Cognition* 12, 739–743.

Wilkinson, F. (1986) Visual texture segmentation in cats. *Behavioural Brain Research* 19, 71–82.

Williams, C.M. and Kramer, E.M. (2010) The advantages of a tapered whisker. *PloS One* 5, e8806.

Wilson, V.J. and Melville Jones, G. (1979) *Mammalian Vestibular Physiology*. Plenum Press, New York and London.

Winslow, C.N. (1938) Observations of dominance-subordination in cats. *The Journal of Genetic Psychology* 52, 425–428.

Wolfe, R. (2001) The social organization of the free-ranging domestic cat (*Felis catus*). PhD dissertation, University of Georgia, Athens, Georgia.

Wolski, D.V.M. (1982) Social behavior of the cat. *Veterinary Clinics of North America: Small Animal Practice* 12, 693–706.

Yamane, A. (1998) Male reproductive tactics and reproductive success of the group living cat (*Felis catus*). *Behavioural Processes* 43, 239–249.

Yamane, A. (1999) Male homosexual mounting in the group-living feral cat (*Felis catus*). *Ethology Ecology and Evolution* 11, 399–406.

Yeon, S.C., Kim, Y.K., Park, S.J., Lee, S.S., Lee, S.Y., Suh, E.H. *et al.* (2011) Differences between vocalization evoked by social stimuli in feral cats and house cats. *Behavioural Processes* 87, 183–189.

Young, J.M., Massa, H.F., Hsu, L. and Trask, B.J. (2010) Extreme variability among mammalian V1R gene families. *Genome Research* 20, 10–18.

Yu, S., Rogers, Q.R. and Morris, J.G. (1997) Absence of a salt (NaCl) preference or appetite in sodium-replete or depleted kittens. *Appetite* 29, 1–10.

Zarzo, M. (2007) The sense of smell: molecular basis of odorant recognition. *Biological Reviews* 82, 455–479.

Zentall, T.R. (2005) Selective and divided attention in animals. *Behavioural Processes* 69, 1–15.

Žernicki, B. (1993) Learning deficits in lab-reared cats. *Acta Neurobiologicae Experimentalis* 53, 231–236.

Zoran, D.L. (2002) The carnivore connection to nutrition in cats. *Journal of the American Veterinary Medical Association* 221, 1559–1567.

Zoran, D.L. and Buffington, C.A. (2011) Effects of nutrition choices and lifestyle changes on the well-being of cats, a carnivore that has moved indoors. *Journal of the American Veterinary Medical Association* 239, 596–606.

Index

Note: **bold** page numbers indicate tables and figures.

PRADSHAW DH
The University of The
9781845939922 = 3 13